Roads, Paths and Bridges

by Logan Waller Page

with an introduction by Roger Chambers

This work contains material that was originally published in 1912.

This publication was created and published for the public benefit, utilizing public funding and is within the Public Domain.

This edition is reprinted for educational purposes and in accordance with all applicable Federal Laws.

Introduction Copyright 2018 by Roger Chambers

Self Reliance Books

Get more historic titles on animal and stock breeding, gardening and old fashioned skills by visiting us at:

http://selfreliancebooks.blogspot.com/

Introduction

I am pleased to present yet another title in our "How To ..." series.

The work is in the Public Domain and is re-printed here in accordance with Federal Laws.

As with all reprinted books of this age that are intended to perfectly reproduce the original edition, considerable pains and effort had to be undertaken to correct fading and sometimes outright damage to existing proofs of this title. At times, this task is quite monumental, requiring an almost total "rebuilding" of some pages from digital proofs of multiple copies. Despite this, imperfections still sometimes exist in the final proof and may detract from the visual appearance of the text.

I hope you enjoy reading this book as much as I enjoyed making it available to readers again.

Roger Chambers

INTRODUCTION

BY THE GENERAL EDITOR

This is the day of the small book. There is much to be done. Time is short. Information is earnestly desired, but it is wanted in compact form, confined directly to the subject in view, authenticated by real knowledge, and, withal, gracefully delivered. It is to fulfill these conditions that the present series has been projected—to lend real assistance to those who are looking about for new tools and fresh ideas.

It is addressed especially to the man and woman at a distance from the libraries, exhibitions, and daily notes of progress, which are the main advantage, to a studious mind, of living in or near a large city. The editor has had in view, especially, the farmer and villager who is striving to make the life of himself and his family broader and brighter, as well as to increase his bank account; and it is therefore in the humane, rather than in a commercial direction, that the Library has been planned.

INTRODUCTION

The average American little needs advice on the conduct of his farm or business; or, if he thinks he does, a large supply of such help in farming and trading as books and periodicals can give, is available to him. But many a man who is well to do and knows how to continue to make money, is ignorant how to spend it in a way to bring to himself, and confer upon his wife and children, those conveniences, comforts and niceties which alone make money worth acquiring and life worth living. He hardly realizes that they are within his reach.

For suggestion and guidance in this direction there is a real call, to which this series is an answer. It proposes to tell its readers how they can make work easier, health more secure, and the home more enjoyable and tenacious of the whole family. No evil in American rural life is so great as the tendency of the young people to leave the farm and the village. The only way to overcome this evil is to make rural life less hard and sordid; more comfortable and attractive. It is to the solving of that problem that these books are addressed. Their central idea is to show how country life may be made

INTRODUCTION

richer in interest, broader in its activities and its outlook, and sweeter to the taste.

To this end men and women who have given each a lifetime of study and thought to his or her speciality, will contribute to the Library, and it is safe to promise that each volume will join with its eminently practical information a still more valuable stimulation of thought.

ERNEST INGERSOLL.

INTRODUCTION

ROAD building is an art based upon a science. In the location and survey of roads, the preparation of plans and estimates, and the selection of materials, the science of engineering plays an important part. A reasonably adequate working knowledge of the art of road building, however, may be acquired by the layman through careful attention to the fundamental principles underlying the building of roads and the methods which have proved sound in practice. For instance, proper drainage will convert an impassable quagmire into a reasonably firm earth road, and a judicious mixing of sand and clay will utilise the good qualities and neutralise the bad qualities of each.

Bridge construction is much more exclusively within the province of the engineer than is road construction, and it is wise economy to incur the expense necessary to secure engineering skill in both road and bridge building. There

INTRODUCTION

are times, however, when the farmer finds it necessary to depend upon his own resources in the building of small bridges, culverts and drains. In such cases a practical knowledge of the simplest theory and practice will often enable him to obtain satisfactory results.

It is the purpose of this book to give in a concise and elementary form the fundamental principles governing the construction of roads, paths and bridges for farm and neighbourhood purposes, and to set forth the details of construction and maintenance so that they may be followed without great difficulty.

A knowledge of the origin and development of road building and the progress of road legislation and administration is not only of interest to the student, but is of real value to every citizen, as it enables him to consider intelligently proposed reforms in road laws and existing systems of administration. The opening chapters deal with this phase of the subject and, in addition, point out a few of the economic aspects of the road question.

CONTENTS

CHAPTER		PAGE
I	History of Road Building	3
II	Road Legislation and Administration	38
III	Locations, Surveys, Plans, Specifications	62
IV	The Earth Road	79
V	The Sand-Clay Road	111
VI	The Gravel Road	124
VII	The Broken Stone Road	134
VIII	The Selection of Materials for Macadam Roads	163
IX	Maintenance and Repair	177
	The Earth Road—The Sand-Clay Road—The Gravel Road—The Macadam Road.	
X	Roadside Treatment	207
XI	Modern Road Problems	215
XII	Paths	229
XIII	Culverts and Bridges	241

ILLUSTRATIONS

A Gravel Road near Savannah, Ga.	*Frontispiece*
	FACING PAGE
Some Ancient Highways	5
Simplon Pass, Switzerland, Pont Napoleon	12
A Paved Street in Pompeii	12
Holland's Highways	21
Thomas Telford	28
John L. McAdam	28
Primitive Methods of Transportation	44
A Toll-house on the National Road	44
Destroyers of Property	49
Economics of Good-Roads Building	53
The Roads and the Schools	60
Examples of Good and Bad Road Location in a Hilly Region	64
Transformation of an Earth Road	85
An Earth Road with Proper Crown	92
An Undrained Prairie Road in Spring	92
Road Machines Hauled by a Traction Engine	101
The Sand-clay Road	108
Three Sorts of Good Roads	129
Constructing a Macadam Road	133

ILLUSTRATIONS

	FACING PAGE
Bad Road-construction	140
Road-making Machinery	144
Effect of Treatment with a Split-log Drag	181
Good and Bad Maintenance	188
Specimen Roads	209
The Automobile and the Road	224
Path of Stone-screenings Beside an Oiled Macadam Road	230
Concrete Culverts and Bridges	244

ROADS, PATHS AND BRIDGES

ROADS, PATHS AND BRIDGES

CHAPTER I

HISTORY OF ROAD BUILDING

SAVAGE man built no roads. His wants were few and of an individual character. When hunger dictated, he sought food in the forests, or in the streams and lakes, and soon came to know the regions where game and fish were most abundant. As time went on he came to know the best and most direct route to his sources of supply, and established for himself definite trails. As he began slowly to mount the ladder of civilisation his habitation became fixed, and communication with his own and other tribes led him to establish more definite routes of travel. Gradually his trails were widened so as to admit the passage of beasts of burden. These widened trails were no doubt our first primitive roads.

At a much later stage in human development

came the wheeled vehicle, and the war chariot was no doubt the precursor of the modern wagon. In history we find the chariot mentioned as early as war itself. The Bible tells of the pursuit of the fleeing Israelites by Pharaoh with 600 picked chariots and a host of others. It is evident from this early general use of the chariot that roads, even in a somewhat modern sense, must have been a necessity at a very early period in our civilisation. The earliest authentic record which we have of stone-surfaced roads is found in Egypt. A little to the east of the Great Pyramids of Ghizeh have been discovered the remains of a great causeway, more than a mile in length. This is supposed to be a portion of the Great Highway built by King Cheops for the purpose of affording a passage across the sands for the transportation of the stone used in the construction of the Great Pyramids. This is no doubt the road on which Herodotus tells us that the Great King employed 100,000 men for a period of ten years. It was built of massive stone blocks, which in places were ten feet thick, and was lined on either side with mausoleums, temples, parks and statues.

Ancient Imperial Highways.—Egypt is not the only land possessing relics of early road building. Babylon, the city of hanging gardens and great walls, at a very early date developed a high state of civilisation, and Semiramis, its great queen, was an enthusiastic road builder. It is at this period that we find what is probably the first use of stone in bridge building, as the two portions of the city were joined by a stone bridge across the Euphrates. Strabo tells us that this wonderful bridge was built of large stone blocks, joined with plates of lead. At this period, more than 2,000 years before Christ, asphalt was used instead of mortar in constructing the vast walls around the city. Commerce flourished and great highways radiated to all the principal cities of the then known world. The highways leading to Susa, Ecbatana and Sardis, are especially mentioned as being lined with travellers and beasts of burden transporting the wealth of commerce. It is said that a highway 400 miles long, and paved with brick set in a mortar of asphaltum, connected Ninevah and Babylon.

The conquering Persians probably learned the art of road building from the Babylonians, and instituted a system of military roads throughout their Empire. The Persians established a military messenger or post service, and at intervals of from 18 to 25 miles stations were established at which riders, whose swiftness Herodotus informs us nothing mortal surpassed, secured fresh mounts. Their speed was estimated at from 60 to 120 miles per day. Strabo states that there were two branches of the great road leading from Babylon to Syria, on one of which only a moderate toll was exacted, and it was, therefore, much more frequented by travellers than the other branch. This is probably the earliest record of the collection of tolls.

As to the details of the construction of these early roads and the system of maintenance in effect, we can judge only by inference; but, as practically all great works of the ancients of which we have definite knowledge were constructed by forced labour, we must assume that their roads and bridges were built and maintained in the same way.

During the time of Solomon two routes from Palestine to Egypt are mentioned as being thronged with travellers, and while no direct mention can be found, several historians inform us that the streets of Jerusalem were paved at least in part during this period. It is plain that exceptionally good roads must have existed to carry on the great commercial trade of the city and to transport the material required for the splendid temple of Solomon.

The ancient Greeks were by no means ignorant of road construction, if we are to judge from the attention bestowed on the subject by the senate of Athens and the governments of Thebes and Lacedæmon. The physical requirements of Greece, however, were such as to call for but few roads. Paved highways are known to have led from Athens to the Piræus and to the sacred shrine at Eleusis. In the ancient city of Thebes the office of telearch, or cleaner of streets, was the lowest office in existence. The ungrateful inhabitants, in order to show their contempt for Epaminondas because of his failure to capture Corinth, elected him to the despised place. But this citizen, whom Cicero declared to have been the greatest man of all times and all nations, held that no man is above the service he can render to the public, and soon through his own efforts Epaminondas raised the office of telearch to be one of the most distinguished favours the people of Thebes could bestow. Later, Thucydides informs us that Archelaus did more for Macedonia than all his predecessors combined, because he promoted the development of the land by making roads, and thus contributed largely to making the interior more accessible.

It was left to the Carthaginians to become instructors to the world in the art of road building. Carthage is given the credit of having demonstrated to the world the strategic and economic value of improved roads. But for a splendid system of highways, which permitted an easy means of communication with all parts of her domains, she never could have reached the heights attained, either in commerce or war. The ready exchange of commerce by land as well as by sea made her able to withstand the terrible drains of long and bloody wars. In spite of the opposition of Athens and all the onslaughts of imperial Rome, and even in spite of the solemn edict, *"Carthago delenda est,"* Carthage continued to stand as an ever-ready menace to Roman supremacy. But the Romans were apt pupils. Ere long they saw at least the military advantage of roads upon which armies and supplies could be moved with celerity—a means, as it were, of increasing the reach of their swords.

There is considerable doubt whether either the Romans or the Carthaginians realised to any large extent the commercial value of their roads. Both built them more as necessary adjuncts to the successful operations of war, offensive as well as defensive, than as avenues for commerce. Good roads were, in fact, the price of existence. They were absolutely necessary for the rapid movement of troops, as well as for providing supplies for large armies. Commerce and exchange followed as a natural result.

We know but little about the Carthaginians as road builders. In so far as they are concerned their art is lost. About the ancient ruins of Carthage are found to-day a few traces of a double road leading to Tunis and occasional traces of a road leading toward Camarat. These and a few ruins are the only visible remains of ancient Carthage, for, in spite of the genius of her commanders, her natural development and great resources, Carthage was unable to withstand the continued onslaughts of her great rival. At last, after nearly 400 years of resistance, offensive as well as defensive, Carthage fell about 146 B. C., and imperial Rome began her mastery of the world.

The Romans as Road Builders.—The Romans are the first systematic road builders of whom we have definite knowledge. The first of their great roads, from Rome to Capua, a distance of 142 Italian miles, was begun by Appius Claudius, about 312 B. C., and is known as the Appian Way, or "The Queen of Roads." This avenue was later extended to Brundisium (Brindisi), or a total of 360 miles, and was probably completed by Julius Cæsar. About 220 B. C. the Flaminian Way was built. This is of special interest because of a stone-arch bridge across the river Nar, 60 miles from Rome. The central arch had a span of 150 feet and a rise of 100 feet, and was pronounced by Addison as the stateliest ruin in Italy. After the completion of the Flaminian Way, road building progressed very rapidly, so that when Rome reached the height of her glory no less than 29 great military roads radiated from her gates. As has already been pointed out, these roads, like those of the Carthaginians, were built largely, if not entirely, for military purposes. They represented the visible efforts which the nation by and through her rulers made for her preservation and the extension of her conquests.

The Roman roads were, as a rule, laid out in approximately straight lines. Mountains, hills and valleys were crossed almost without any regard to topography. Hills were cut through and deep ravines filled in. Although these roads remain to some extent even to-day as splendid monuments of their builder's art (the Appian Way is said to have been in good repair 800 years after it was built), we can hardly credit these people with having intentionally built for the ages. More than likely the ponderous construction they adopted was that which they, to the best of their ability, believed necessary for a reasonably permanent and satisfactory road.

The extent of the Roman road system is astounding. Every conquered province was soon traversed in all directions with connecting roads. Of the narrow paths, three to six feet wide, found in conquered Gaul, no less than 13,000 miles are said to have been improved. In Britain, the road improvement is estimated to have been at least 2,500 miles. Across the Alps, through Gaul to Spain, Austria and the regions of the Danube, led the great military roads. Nor were the countries beyond the seas ignored. Straight to the water's edge led the road from Rome, and then on the shore beyond was the continuation. England, Sicily, Africa, and even Asia, all bear witness of the wonderful energy which strove to bind firmly every member of the great empire into a living whole.

Nor was this energy directed exclusively toward imperial progress and the building of roads for the movement of legions, or to satiate an empire with the luxuries of remote countries. If not at first, at least later in her development, Rome saw in her roads value other than military, for in the reign of Augustus we find a seemingly well-devised system of crossroads leading to and connecting villages and even farms with the great military roads. The roads were no longer exclusively military, but were also filling the domestic needs of the farmers.

The construction proper of the best type consisted of four courses. The *statumen* or foundation was composed of large flat stones bedded in mortar. On this foundation was placed a layer of hand rock laid in lime or cement mortar, known as the *rudus*. The nucleus consisted of small stones, gravel, or pieces of brick and broken tile, laid in mortar. On top of this was placed the *summa crusta,* or wearing surface, of large, flat, closely bedded rocks, making the total depth of the road about three feet. Under present conditions and prices of labour, even aided by all our modern machinery, such roads would probably cost from $50,000 to $200,000 per mile, and though the Roman roads were built in part by slave labour and the spoils of war, they represent an enormous outlay.

The Roman roads, though solid enough to bear the heaviest loads and durable beyond question, still lacked many of the essentials of a good road. The steep grades and their excessive hardness made them very wearisome to travellers, and Horace informs us that "they were less fatiguing to those who travelled slowly." Even on these roads, which we would think nothing could injure, we find that the weight and nature of the traffic were closely regulated by statutes which were rigidly enforced.

The highway legislation of the Romans forms the basis for our present road laws. By the Roman law, the use of the roads was for the public. The roads could be the property of no individual, while the emperors or other chief magistrates were their conservators. The majority of the main highways were built by contract at the public expense. Their maintenance was in part by the labour of soldiers, convicts or slaves, or by an enforced service which, in some instances, took the form of taxation. In whatever form the maintenance was made, it was at the expense of the district through which the road passed. Tolls as a means of repairing highways appear to have been seldom resorted to. Some of the Roman roads were constructed through the private munificence of her emperors or other great personages ambitious of popularity, or with the spoils of war brought home by successful generals. The supervision of the roads was entrusted to men of the highest rank. Augustus himself seems to have made those about Rome his special care. The crossroads or vicinal roads were committed to the charge of the local magistrates, and, as a rule, maintained by compulsory labour, or the contributions of the whole neighbourhood, although occasionally a portion of a road was assigned to some landowner to maintain at his own cost.

Roads in Ancient Peru and Mexico.—The countries of the Mediterranean are not the only ones which have developed systems of roads. The ancient civilisations of Mexico and Peru had roads which we are told were in some instances not inferior to those of the Romans. The Incas of Peru had a magnificent system, extending to every part of their vast empire, but, as far as we can learn, they were largely, if not entirely, built to accelerate the movement of troops and supplies. So well did they also serve an economic purpose, however, that, prior to the advent of the white man, such a thing as famine was unknown in Peru. At stated intervals along the main road were tambus or caravansaries and storehouses, where provisions were collected for the soldiery, so that the passage of troops never entailed any additional hardship on the people along the way. These roads also served as a system of post roads for the rapid transmission of government dispatches. About every five miles were stationed runners selected for their speed, endurance and reliability, who acted as carriers.

The most magnificent of the Peruvian roads was the great mountain highway between the two capitals, Quito and Cuzco, of which only a few fragments remain to-day. The younger Pizarro on first obtaining sight of it exclaimed: "Nothing in Christendom equals the magnificence of this road across the Sierra." Humboldt, who viewed the remains in the beginning of the 19th century, said: "Nothing I have seen of the remains of Roman roads in Italy, in the south of France and in Spain, was more imposing than the works of the ancient Peruvians, which were, moreover, situated at an elevation of more than 13,000 feet above the level of the sea;" while Cieza, who saw the road about 1540, compares it to the Roman road in Spain at that time, which was known as the "Silver Road." The length of this magnificent road is variously estimated at from 1,500 to 2,000 miles, or five or six times the length of the completed Appian Way. The breadth of the road scarcely exceeded 20 feet. Near Cuzco we are told that there was a stream of water and shade trees on either side, while stone pillars at regular intervals, similar to European milestones, marked the distances. Besides the regular tambus there were also at frequent intervals smaller buildings exclusively for the accommodation of travellers.

The construction was ingeniously varied with the requirements of the region traversed. Portions greatly exposed to destructive agencies were paved with massive blocks of well-cut stone, sometimes as much as 10 feet wide. Other regions were paved with a substance not unlike bituminous macadam, which Humboldt says "time has made harder than the rock itself." Wide rivers in deep canyons were crossed on suspension bridges composed of fibre, some of them being more than 200 feet in span. In the adaptation of roads to natural conditions, the Peruvians were superior to the Romans. Instead of clinging to the straight line, the Peruvian roads were adapted to the topography of the country. No avoidable ascents were made. Unavoidable precipices were scaled by means of steps, and since wheeled vehicles were unknown and the llama was the universal beast of burden, this was no serious obstacle to travel.

Mediæval Neglect and Its Consequences.—With the fall of the Roman Empire, its magnificent system of roads passed into disuse and neglect. With Charlemagne came a slight revival, but the economic and political conditions were such as to make this impulse of but short duration, and the country soon lapsed into feudalism. Each little community depended upon itself for the necessities of life. Commerce was practically abandoned. The roads came to be looked upon with dread, and as being simply avenues upon which the robber barons might at will swoop down for plunder and rapine. Seclusion and inaccessibility came to be considered as in a measure essential to safety. In many places the roads were torn up and destroyed in order to prevent the easy ingress of robbers and marauders. The little travel done was on foot or horseback, along narrow paths or trails.

At this period the old Roman highways had

sunk into the marshes or been overgrown by forests, and such other roads or paths as existed are described as having been "in a state of nature, or worse." A road was simply a right of way, an unimproved path from one hamlet to another. Almost all goods were transported by packhorses. In some parts of the country wheeled vehicles were entirely unknown. If the road was extraordinarily bad, the traveller left it and travelled in the adjoining field or wood.

At the end of the eleventh century came the first crusade, which was followed during the next two hundred years by seven other similar movements of greater or less magnitude. Though they failed of their original object, the crusades were of immense value to the whole of Europe, in that they promoted intercourse between the nations, awakened them from their lethargy, and stimulated commerce and the dissemination of knowledge. In order that it might be possible to move armies toward Syria, towns made grants, and kings and popes contributed money and issued edicts for the improvement of highways.

Revival of Road-Making in Europe.—Commencing with Louis XII in 1508, successive kings appointed road overseers for the kingdom of France. These officers bore various titles at different times. In general, however, they were charged with the duty of inspecting the "King's Highways" and repairing them. Even at this early time the French roads seem to have been divided into two classes: viz., those main lines between cities, known as the King's Highways, and the minor crossroads, under the charge of the nobles through whose territory they ran. It appears to have been about the beginning of the sixteenth century that some systematic repairs on roads were begun, though for another 150 years nothing more than filling up the worst holes was attempted.

In the reign of Henry IV France emerged from mediævalism. Sully, who was appointed Comptroller of Finance in 1597, was also made Grand Voyer or Great Way-Warden of France. Owing to the interest this first of the great ministers took in agriculture, a beginning was made in the improvement of the roads through the rural districts. Up to that time France had not had a smooth, hard roadway since the Roman supremacy. A few, such as the road from Paris to Orleans, had a roughly paved causeway in the centre, but such was the condition that all travel was necessarily by horseback. In the winter it was almost impassable by any means. Under Sully, and later under Richelieu, slow progress was made in bettering the condition of the roads. At the beginning of the seventeenth century broken stone began to be applied to the roads for their improvement. In 1661 Colbert was appointed Comptroller of Finance, which office carried with it the superintendence of highways. During his ministry 15,000 miles of hard road were constructed. Such a great work was not accomplished without a corresponding hardship to the people. The old feudal institution of the corvée was used to an extent hitherto unknown. The peasantry were taken long distances from their homes and kept at work on the roads even during seeding and harvest times. Rioting and insurrection were provoked in various parts of the kingdom, and the condition of the people became almost unendurable. This system prevailed with the utmost rigour until 1774, when Turgot, who was then Minister of Finance, relieved it of its worst features, but it was not finally abolished until 1787, when the nation was on the brink of revolution.

The present magnificent road system of France was really founded by Napoleon. He built many roads through the empire, among others the famous road over the Simplon Pass in Switzerland, which was commenced in 1800, and required six years for completion. The road work of France was systematized and placed in the hands of a competent and permanent body of engineers, and, in order to raise funds for the continuance of this work, Napoleon attempted to establish a toll system on the best roads. Owing to the determined opposition of the people, however, the idea was soon abandoned. From this time on road improvement has been extremely popular in France and the people have willingly submitted to the necessary taxation.

In 1775 the great French engineer Trésaguet published his first treatise on broken-stone roads. Too much credit can not be given to the work of Trésaguet. He was the real father of modern road building and maintenance, as his work preceded that of McAdam and Telford by about forty years.

Pierre-Marie-Jerome Trésaguet was born at Nevers, France, in 1716, and was made Chief Engineer in the District of Limoges, July 22, 1764, at a salary of 2,400 francs ($480) per annum. His duties consisted in supervising the construction and maintenance of bridges and highways in the District, and it was here that he conducted his most important work. On April 19, 1775, he was made Inspector General, at a salary of 3,600 francs ($720) per annum, and 2,400 francs ($480) for travelling expenses. Later on in the same year, he was made a member of the Commission of Inspectors General of France, but he continued his duties at Limoges for about two years. He became very famous, and in 1785 he endeavoured to introduce his system in Paris. About this time, however, he fell sick, and his great worth was recognised by his being awarded a pension of 3,000 francs ($600). During the French Revolution this pension was reduced, and, at the age of 80 years, he found himself reduced to the direst poverty. In that year the commune of Paris was called on to give this eminent engineer three pounds of meat daily. He died in the same year.

In his report to the Assembly of Bridges and Highways in 1775, Trésaguet pointed out that, although the ancient highways were of a thickness of eighteen inches in the middle and twelve inches at the sides, in six months they were cut with deep ruts, because of the lack of maintenance. He suggested reducing the depth of material in the centre to ten inches, and that the sides be sloped at an angle of about 20 degrees. Trésaguet laid great stress on systematic, continuous maintenance as against intermittent and irregular repairs. It was he who organised the canton, or patrol system, which has made the French roads the most superb in the world.

Development of Road Building in England.— The first record of road legislation in England bears the date of 1285, and provides that the trees and bushes on both sides of the road for a distance of 200 feet shall be cut away to prevent robbers from lurking therein and rushing upon their victims unawares. It further provides that when a road is worn deep another shall be laid out alongside.

This latter provision was slightly modified by Henry VIII about 250 years later, who provided that "two justices of the peace and 12 other men of

HISTORY OF ROAD BUILDING

wisdom and discretion shall choose fresh routes when the old ones are worn out." In 1346 Edward III authorised the first toll to be levied for the repair of roads. This commission was granted to the Master of the Hospital of St. Giles and to John Holborn, authorising these parties to levy toll on vehicles passing on the road leading from the hospital to the Old Temple of London, and also on an adjoining road called the Portal. In 1523 Parliament passed its first act relative to the repair of roads, but it was not until near the middle of the 18th century that highway legislation became active.

The condition of the streets and roads of England was indeed deplorable during this time. Street pavements developed somewhat earlier than rural roads, but even their improvement was extremely slow. As late as 1190, we are told that a wind-storm unroofed the church of St. Mary-le-Bow, Cheapside, London, and four pillars 26 feet long, falling vertically in the street, sank by their own weight, so that only four feet remained above the mud. These were certainly not very inviting streets for either pleasure or business. Still, it was not until 1532 that the first statute for paving in London was recorded.

In this modest act the streets are described

as "very foul and full of pits and sloughs, so as to be mighty perilous and noyous as well for all the King's subjects on horseback as on foot with carriages." Nor did this condition change rapidly. Writing in the year 1770 Mr. Arthur Young, after a six months' tour throughout northern England, says of the turnpike to Wigan:

"I know in the whole range of language no terms sufficiently expressive to describe this infernal road. Let me seriously caution all travellers who may accidentally propose to travel this terrible country to avoid it, as they would the devil, for a hundred to one they break their necks or their limbs by overthrow or breakings down. They will meet with ruts, which I actually measured, four feet deep, floating with mud only from a wet summer; what then must it be in winter? The only mending it receives is tumbling some loose stones into the worst holes, which serves no other purpose than jolting a carriage in the most intolerable manner. These are not only opinions, but facts; for I actually passed three carts broken down in those 18 miles of execrable memory."

Still later Lord Macaulay informs us that the roads were so bad that in places the crops were allowed to rot in the fields, while only a

HISTORY OF ROAD BUILDING 27

few miles away people were actually dying of starvation. With such roads, a few miles were a more effective barrier than the oceans are to-day. Not only was commerce practically impossible, but even the news of dearth or plenty could travel but slowly.

The legislative effort to better the condition of the English roads expressed itself in a comprehensive system of turnpike acts. It is estimated that in 1838 about 1,100 of these turnpike trusts were in existence throughout the kingdom. They proved of little permanent value, however. Not only their construction, but also maintenance was often defective. The cost of collecting the tolls often nearly equalled the income, leaving little or nothing for maintenance. In 1871 the census showed that 5,000 persons in England and Scotland were engaged in merely collecting tolls. In 1857 Ireland freed herself from toll gates, and tolls were abolished in England by act of Parliament, passed in 1878.

McAdam and Telford.—No historical sketch of the highways of England would be complete without at least a mention of the two great engineers, John Loudon McAdam and Thomas Telford.

John Loudon McAdam was born at Ayr, Scotland, September 21, 1756, and spent his boyhood in New York. He returned to Scotland in 1783, and from that time until 1798, he was Trustee of Roads and Deputy Road Lieutenant of Ayrshire. In 1798 he moved to England and was made Superintendent of Roads of the Bristol District in 1815. He made a most exhaustive investigation of roads in England, and he is said to have travelled 30,000 miles and spent more than five years and £5,000 in investigating the English roads. He made a report in 1811 to a committee of the House of Commons, outlining his system. In 1827 he was made General Surveyor of the Metropolitan Roads, and in recognition of his success in improving them, he received a grant of £10,000 from the British Government. His methods are set forth in the chapter on macadam roads in some detail. McAdam died at Moffat, in Dumfriesshire, November 26, 1836. He wrote two books, which have become classics in road-building. They are: "A Practical Essay on the Scientific Repair and Preservation of Roads" (1819), and "Present State of Road-Making" (1820).

Thomas Telford, a civil engineer, was born August 9, 1757, at Eskdale, Dumfriesshire, and was the son of a shepherd. At the age of fifteen he was apprenticed to a stone mason at Langholme, where he had an opportunity to gain an acquaintance with Latin, French and German. As a young man, he was fond of writing poems, a number of which were published, although they were of comparatively little value. In 1780 he went to Edinburgh, where he was employed in the erection of houses, and occupied much of his time in learning architectural drawing. In 1782 he went to London and found employment in the erection of Somerset House. This was followed by other work of a similar character, which eventually resulted in his appointment as Surveyor of Public Works for the county of Sallop. His first bridge was finished in 1792. Later on he was employed in the construction of some of the most important canals in Great Britain, and was consulted in 1806 by the king of Sweden regarding the construction of the great Gotha Canal, for which his plans were adopted. In 1803 he was appointed engineer for the construction of 920 miles of road in Scotland, most of which was through difficult country. He also built a system of roads in the most inaccessible parts of Wales, where he built a most magnificent suspension bridge across the Menai Straits. He also built an important road for the Austrian Government from Warsaw to Brest. He did very important work in the improvement of harbors, and was generally looked upon as one of the most eminent engineers of his time. His great work in the building of roads and bridges has given him the most lasting fame, and a type of road which is now quite frequently built on marshy or unstable ground is known as the Telford road.

Early Road Work in the United States.— America, as an abode of the white man, was still young when she entered the field of road building and road legislation. The first American road law was passed by the House of Burgesses of Virginia in 1632. So far as can be ascertained, the first American road built by white men was at Jamestown a few years later. We can imagine the conditions somewhat, both as to means of communication and transportation, when we learn that in 1625, when the British Crown took over Virginia from the London Company, the inventory revealed the interesting fact that the Governor alone had a horse.

In the North the so-called New England Path, between Boston and Plymouth, was begun in 1639. In the province of New York, regulations for road building were passed in 1664, and two years later the first Maryland road law came into existence. Pennsylvania followed some years later (1692) with a road act placing the control of the highways in the hands of the townships which, however, was amended eight years later, whereby the control was given over to the counties. To Pennsylvania is also given the credit of the first important macadam road built in America—the Lancaster turnpike from Lancaster to Philadelphia—which was constructed in 1794. Portions of this road are still operated as a toll road.

The extent and character of these early roads may perhaps be judged more clearly from the state of the postal service. It was not until 1673 that a post service was established between New York and Boston, and three days were required for the trip. Twenty-two years later, in 1695, letters were forwarded only eight times a year from the Potomac to Philadelphia. In 1717 mail from Boston to Williamsburg, Virginia, was delivered every four weeks in summer and every eight weeks in winter, and as late as 1790, the number of post-offices in the United States numbered only 75.

Thus, at the time of the organisation of our Government, highway construction can scarcely be said to have begun. The few roads, if they may be dignified by such a name, were mostly the result of chance—mere trails which had gradually been widened to admit the passage of vehicles, but were usually almost impassable during long periods. Systematic organisation for either construction or maintenance did not exist, and each little community was left to cope with the problem as best it could.

Civilisation was rapidly pushing toward the great West, however, and the need of roads became imperative.

The first wagons crossed the Alleghanies within two years after the close of the War of the Revolution. The need for better roads had now become so strong that private capital was attracted and numerous toll roads were constructed throughout the different States. By 1828 nearly 2,380 miles of these

HISTORY OF ROAD BUILDING 33

roads had been constructed in Pennsylvania alone, at a cost of $8,431,000. Few, if any, of the turnpikes returned sufficient dividends to make them a profitable investment, as one of the chief drawbacks was the high cost of maintaining toll gates and collecting the tolls. Nor did the turnpikes suffice to fill the demands of the time. In 1821 the cost of transporting a barrel of mackerel from Philadelphia to Somerset was $8 per hundred pounds, and from Philadelphia to Pittsburg the rate was $11, or 70 cents per ton-mile. Not until about 1865 were the railroads of sufficient extent to make themselves felt as considerable factors in the wholesale reduction of long-distance freight rates. By this time the charge for hauling freight from Baltimore to Wheeling on the turnpike road was reduced to 17 cents per ton-mile.

Our national legislators early recognised the need of adequate means of communication and transportation, and after a lengthy debate an act was passed in 1806 providing for the building of a great highway from the Atlantic to the Mississippi. Beginning at Cumberland, Md., on the Potomac, this great highway passed through the States of Maryland, Pennsylvania, Ohio and Indiana, westward to the Wabash and the Mississippi. For thirty-two years the Government struggled with this great enterprise until finally the appropriations ceased altogether in 1838, and the work was discontinued after an expenditure of $6,824,919.33 appropriated by Congress.

President Monroe once vetoed the appropriations for the National Turnpike, as well as a bill introduced by John C. Calhoun, providing for setting aside the dividends from the National banks for road purposes. Revenues derived from the sale of public lands, however, continued to be set aside by Congress for aid in road construction. Between 1811 and 1845 Louisiana, Indiana, Mississippi, Illinois, Missouri and Iowa were aided in this way to the extent of about $5,000,000. Between 1854 and the beginning of the Civil War Congress appropriated in all something like $1,600,000, which was expended chiefly on roads within the territories. Thus, up to 1861, the National Government had assisted in road building throughout the Nation to the extent of about $14,000,000.

Since 1861 the National Government has rendered aid to road building only in an educational sense. For a time following the war the immense debt incurred made appropriations from the National treasury almost out of the question, and besides, the idea had become quite prevalent that the railroads had lessened the need as well as value of improved

HISTORY OF ROAD BUILDING 35

roads. It required some time for the country to discover the error and it was not until the advent of the bicycle that the good-roads movement awoke from its lethargy.

Government Aid.—An office of road inquiry was established under an act of Congress, approved March 3, 1893, making an appropriation of $10,000 to the Department of Agriculture for making inquiries in regard to systems of road management throughout the United States, and for making investigations in regard to the best methods of road-making, preparing didactic publications on this subject, and assisting the agricultural colleges and experiment stations in disseminating information.

The work of the office was at first of necessity very limited. In 1897 the construction of short sections of sample roads under the supervision of skilled road builders was begun in a small way in co-operation with the various agricultural experiment stations. In December, 1900, a laboratory for testing the physical qualities of different road-building materials was added. Two years later the annual appropriation was increased to $20,000, and provision was made for the investigation of the chemical and physical character of road materials. The language of the appropriation bills has remained practically unchanged up to the present time, except that the name of the office was changed from Public Road Inquiries to the Office of Public Roads, and a statutory organisation provided in the agricultural bill, approved March 3, 1905.

From this modest beginning the work of the office has gone along in a careful and conservative manner.

Local communities can easily avail themselves of the assistance granted by the United States Office of Public Roads. It is necessary only for the local authorities having jurisdiction over the roads to make application either to the Secretary of Agriculture or the Director of the Office of Public Roads for the assignment of an engineer or expert to investigate local conditions with reference to the roads, and to give such advice and instruction as may be necessary. The salary, and in most cases the expenses, of such an engineer are paid by the National Government, and hence his services are free to communities. The road-material laboratories of the Office of Public Roads make tests to determine the relative value of road material, without cost to any citizen of the

United States who will take the trouble to write to the Office and obtain the necessary forms and shipping blanks for submitting samples of material. The only expense to be borne by the private individual is the transportation charge on the material to the Office.

Within recent years the investigative work of the Office of Public Roads has attracted world-wide attention, and the testing laboratories are looked upon as equal to, if not superior to, any in existence.

CHAPTER II

ROAD LEGISLATION AND ADMINISTRATION

While in its strict interpretation a principle is defined as a fundamental truth or doctrine, we are at liberty, in dealing with this subject, to regard a policy which has been followed by many agencies over long periods of time and with successful results as being fundamental, and, therefore, to be considered in the light of a principle.

Applying this reasoning to the information afforded by the histories of the road systems of all countries, it becomes evident that one of the features common to all of the successful road systems of history is centralisation of authority and responsibilities. The most striking example illustrating the power of centralisation is afforded by the splendid roads of Rome. No sooner had the power of the imperial city crumbled away and the management of her splendid roads passed into the hands of many nations than the roads began to deteriorate, because of the utter absence of attention. The most conspicuous example in modern times of an efficient system of roads well constructed, maintained and administered is the road system of France. Beginning with the humble patrolman, the system provides definite lines of authority through various grades upward to the Inspector-General at Paris, whose guiding hand directs the whole organisation, prevents duplication of effort and co-ordinates all efforts and all accomplishments. In the United States the States which have made most progress in the actual improvement of the public roads are those which have in some degree centralised the construction and care of the roads in a State official or officials.

Until comparatively recent years most of the States of the Union have followed a policy directly opposed to the policy of centralisation. The laws all provided, and in many States continue to provide, for a large number of officials, each having a very limited territory under his control, and each being, in a measure, independent of any direct supervision. This policy of extreme localisation has, by its very failure to produce adequate results, confirmed the wisdom of a centralised system.

In the light of this evidence it seems clearly demonstrated that each State should provide a centralised direction of its road work, and, pending that time, each county should, as far as possible, centralise the control of its road work in a competent official, and, carrying this reasoning still further, it is safe to say that each township could with profit place its work under the direction of a competent employé or official rather than to depend upon a number of officials whose authority is not defined and whose duties overlap.

A second important point which has been brought out clearly by the experience of all nations, and which has been emphasised most strongly since the development of modern traffic conditions is the necessity for special knowledge and skill on the part of the men who actually build and maintain the roads. It is a curious fact that, although the public road is conceded to be so important to humanity as to be classed with the home, the church, and the school, and although its condition directly affects the welfare of all who are called upon to traverse it, and indirectly all who are dependent upon the products which are transported upon it, few people give more than a passing thought to the methods by which the road is built and maintained, while it is an inborn conviction on the part of nine men out of ten that they are thoroughly competent to say how a road should be built and maintained. If the general public would consider that to build a successful road a suitable location must be found, grades reduced where necessary, a drainage system provided, suitable material selected, foundation and surface arranged with great exactness, culverts and bridges constructed, and, to meet modern traffic conditions, the whole subject of bituminous and other special binders dealt with from the standpoint of the expert, there would be more inclination to employ for such work men who make a profession of highway engineering. It is very difficult to find in any of the road laws of this country, except those that provide for State Highway departments, any requirement that the officials having charge of the road work shall possess any special qualifications. Hence the enormous waste of public funds through ignorance of correct methods, as well as from the lack of centralised authority. Skilled supervision is an essential in road work, and should be considered a fundamental requisite applicable at all times, under all conditions, and by all units of government. It is just as necessary for the township to employ a man with a knowledge of road building rather than one who has no knowledge of the subject, as it is for the State to require its highway engineer to be competent and experienced.

Roads Belong to the Public.—That the roads belong to the public and that their use and control should remain with the public is a principle recognised by Rome in the management of her great system of highways, and which has persisted in all of the civilised countries of Europe, and has finally been recognised throughout the United States, in spite of costly and elaborate experiments with the toll-road system, particularly in England and in this country. The Romans never approved the plan of giving over any of the public roads to the control of private individuals or companies, and few, if any, tolls were ever collected on Roman roads. The English Parliament in the latter part of the eighteenth century passed innumerable turnpike acts, and for a good many years during the eighteenth and nineteenth centuries the toll-road system was supreme in England. It broke down under the fierce resentment of the public, and because it was costly and clumsy in operation. The cost of collecting the tolls was totally out of proportion to the amount actually spent in maintenance. Toll roads were abolished finally in Great Britain in 1878. In this country the beginning of the nineteenth century witnessed very great activity in the building of pikes or toll roads. Here, however, as in other countries, the experiment proved unprofitable and contrary to the public welfare, with the result that this system has been gradually abandoned, until at the present time the toll road is decidedly the exception rather than the rule in the United States. The toll system is fundamentally wrong because it places under private control that which must of necessity be a public utility; it places the burden of taxation solely upon the users of the road, and leaves untaxed those who benefit materially from the improvement of the road, although having no occasion to make use of it for travel. An example of this is to be found in non-resident owners of tracts of land abutting the road and increasing in value by reason of the improvement. Finally the toll system is unprofitable to the stockholder and excessively burdensome to the traveller, because of the great cost of collecting the tolls and conducting the system, which makes the dividend low to the stockholder and the rate high to the traveller.

Personal Service on Roads Inadvisable.—From ancient times the practice has been general among all nations to compel personal services on the roads, or to accept personal services in lieu of a cash tax. Under the ancient despotic monarchies slave labour was largely used, and under the Bourbons of France the peasants were compelled to contribute a number of days' work on the public roads. Modern standards of humanity discountenance these rigorous methods, but they exist in another form through what is known as the statute-labour system.

Under this system the laws require that each able-bodied citizen perform a given number of days' service upon the road, or commute this labour tax in cash, while in many States of the Union, even the taxes that are payable in cash may be paid in labour at a given rate per day.

This inadequate system is entirely out of harmony with modern business practice and modern governmental policies. It provides untrained labourers who are not amenable to discipline and who render their services grudgingly and in as scant a measure as possible. They are at the same time employers and employés, because it is by their votes that the road officials are kept in power. In consequence they dominate their leaders and the results which they accomplish are almost negligible. Hence it has come to be essential to efficiency in the administration of our public roads that all road taxes be paid in cash, so that regular employés may be obtained, who may be required to give a full and honest day's work, who may acquire the skill essential to efficiency, and who may be answerable to reasonable discipline. We can hardly consider it, however, a maxim that road taxes should be paid in cash, regardless of all conditions. It has been claimed that in some of the Southern States it is impossible to collect cash taxes, and the only recourse is to compel personal services on the part of a large element of the population. It must be understood that such examples constitute exceptions, and that recourse to the payment of road taxes in labour is justified only in extreme cases.

The problem of what to do with the convicts and other offenders against society has been one which has vexed the students of sociology for centuries, and it is now universally conceded that idleness is extremely detrimental to the prisoner, and by reason of his unproductiveness, burdensome to the public. Outdoor productive labour is conceded to be beneficial to the prisoner mentally, morally, and physically, and to make possible a return to society for its outlay. As to the character of work which should be performed by the convict, it is reasonable to assert that as the prisoner has offended against the public, his labour should be for the benefit of the public, and directed toward public improvements. In this way, he is not only promoting the public welfare, but he is also entering into competition to the least possible degree with honest free labour. The volume of public improvements is necessarily limited, and comprises, among other improvements, the construction of roads and the preparation of road materials. Certainly no public improvement upon which the convict can be employed will yield a greater amount of benefit to the public than the improvement of the roads, and wherever this form of labour is applicable, it should be employed. In the South excellent results have been obtained by using convicts in actual construction of roads. In some other States the convicts have been employed in stockades in the preparation of road materials. Conditions are such in some of the States as to make the wisdom of using convicts in this way questionable, but the plan should not be rejected without the most thorough consideration.

Roads More Than Local Institutions.—By reason of the many inventions of modern times which have tended to shorten distance and time, which have enormously increased manufactures, and which have made possible the concentration of a large percentage of our population in cities, and because of the growth of education, the general dissemination of learning, and the broader field of knowledge afforded to the people of civilised nations in the present day, the isolation of local communities has been largely superseded by the intercommunication and interdependence which link together communities hundreds of miles apart. The road is no longer a merely local institution, for over it must be transported the food products which are necessary for the existence of the city dwellers, and the manufactured products which come from the city to the country dwellers. This road may be traversed by the automobilists from other neighbourhoods and other States, and by the transient guests of the summer hotels and resorts. The condition of this road affects the welfare, not only of the people who live near it, but of all those other classes of people who have occasion to buy the products of the surrounding country, or to sell to the inhabitants, or to make use of the road as transients. This condition has given increasing importance to the maxim that all who share the benefits of road improvement should share the burdens incident to such improvement. This maxim has found concurrent expression in the establishment of State highway departments and the appropriation of State funds to aid in the improvement of the main travelled roads. State aid is justified, not only on the ground that it distributes the burdens in proportion to the benefits, but also because it provides a centralisation of authority, skilled supervision, and the public control already referred to as essentials, and as a result of these factors, economy, co-ordination and tangible results in the way of construction and effective maintenance follow.

Importance of Systematic Maintenance.—Effective maintenance of the roads is rather a result than a system and, if the other essentials, namely, centralised skilled supervision, cash taxation, public control, and the utilisation of convict labour, be adopted, it is probable that effective maintenance will follow. It is well to state here, however, that almost without exception no provision has been made in the United States for the maintenance of roads, even those which are most perfectly constructed and which would, therefore, seem to justify some outlay for the maintenance of their high state of efficiency. It is just as careless and unwise to leave a good road uncared for as it is to leave a well-constructed building to the mercy of the elements and depredations of the public. The strongest feature of the French road system is the constant care of the roads which have already been constructed. The whole system of main roads is divided into short sections of from 2½ to 5 miles approximately, and each section is in charge of a patrolman, who gives his entire time to the road, repairing slight defects as soon as they occur, keeping the ditches open, trimming the trees and bushes, removing dust and deposits of sand and earth after heavy rains, and, when ordinary work is impossible, he prepares stone and transports it to where it is needed. In order to facilitate this repair work quantities of crushed stone and gravel are placed at convenient intervals along the road, while to meet the expense of this maintenance annual appropriations are made, based upon careful estimates by the engineers in charge. Maintenance, to be effective, must be systematic or in accordance with some definite plan or purpose, and must be continuous, instead of at long intervals, as we practise it in this country.

Financing Public Roads.—The methods of financing road improvement constitute a very important part of the subject. Eliminating as unwise and impracticable the toll system and, except in extreme cases, payment of road taxes in labour, it follows that there are only four ways of obtaining revenues for improving the roads, namely, a cash property tax, a poll tax, a bond issue, and a State appropriation, the latter of which may be derived from one or many sources. There may be special methods of obtaining revenue which are applicable only in special cases, such, for example, as private subscription, the sale of public property, the appropriation of certain license taxes, etc., but as a general proposition the four sources of revenue already named must be depended upon. It is manifest that State aid cannot be given in sufficient amount to meet the needs of the respective counties. Poll taxes may not be available, as the law may provide that they be expended for other than road improvement. Even where this form of revenue is available, it is usually necessary to supplement it with some other form of revenue. This brings us to the consideration of the relative advantages of making only such improvements as may be possible by means of an annual cash tax, and the making of improvements on a large scale by means of a bond issue. It may be said in favour of bond issues that they bring immediate returns in the form of improved roads and in such amount as to enable a large proportion of the population to enjoy the benefits of this improved medium of transportation without having to wait a long period of years. The improvement necessarily develops the resources of the locality more quickly and thereby increases wealth. The cost per mile of road is lowered by reason of the magnitude of the enterprise, while the cost of maintenance is materially decreased because it is easier to maintain a long stretch of improved road connecting two communities than it is to maintain short sections of improved roads, the ends of which ravel or disintegrate more quickly because of the fact that the remainder of the road extending from each end of the improved section remains unimproved. A bond issue generally places upon the next generation a portion of the burden, but this is contended to be equitable by reason of the fact that the wealth thus developed inures to the benefit of the generation called upon to bear a portion of the burden.

A bond issue should never be considered a wise undertaking simply because it is a bond issue, nor should it be considered unwise for the same reason. The needs of the community in the way of improved roads, the financial condition of the community, the necessary outlay to obtain this improvement, and the probable resultant benefits, compared with the resultant burdens, should always be considered. The Office of Public Roads of the United States Department of Agriculture maintains a corps of engineers who are qualified to examine local conditions intelligently, and recommend a plan of improvement and outlay commensurate with the needs and the ability of the localities which they are called upon to advise. The services of these engineers are given free by the Government. Moreover, in many of the States having State highway departments, assistance of this kind can be secured without cost. It is recommended, therefore, that the expenditure of large sums of money be based upon such careful and conservative advice.

Road Economics.—Economics deal with that phase of the road subject which has to do with the relation between the outlay for road improvement and the returns in the form of benefits. The question to be considered in all cases is not whether the outlay is large or whether the benefit is indirect, but whether the resultant benefit, either direct or indirect, is greater than the outlay. The amount to be expended can easily be ascertained by means of specific designation and conservative estimates of the improvements proposed to be made. The methods of providing the necessary revenues can be determined and the necessary administrative requirements made along the lines indicated in the preceding paragraphs. It remains, therefore, to consider what benefits may be expected to arise from the proposed improvements, and what distribution of the improvements will afford the greatest amount of benefit to the greatest number of people.

ROAD LEGISLATION

In the first place the improvement of the road may be expected to lower the cost of hauling, greatly increase facilities for transportation, and add to the comfort of those who must use the road for these purposes. As a basis for considering this phase of the subject it may be stated that in 1906 the Bureau of Statistics of the Department of Agriculture obtained data which indicated the average cost of hauling to be 23 cents per ton-mile, and the average haul $9\frac{9}{10}$ miles. In the report issued by that Bureau it was stated that although ocean rates were higher than usual during the year 1905–1906, the mean charge for carrying wheat by regular steamship lines from New York to Liverpool was only $3\frac{3}{10}$ cents per bushel, or $1\frac{9}{10}$ cents less than it costs the farmer to haul his wheat $9\frac{9}{10}$ miles at 19 cents per ton-mile, from his farm to a neighbouring railroad station. Moreover, the cost for hauling wheat is less than the general average for all products. It is generally known that the load which two horses can draw on a smooth, hard road is double and sometimes treble the load which they can draw on an earth road. Engineers have made investigations on this point which indicate that the difference in cost of hauling upon broken-stone roads, dry and in good condition, and an earth road containing ruts and mud, is the difference between 8 cents per ton-mile and 39 cents per ton-mile. Since the introduction of motor vehicles this cost has been still further lowered, and a special demonstration of motor trucks in California yielded a rate of about 2½ cents per ton-mile. It is manifestly impracticable to improve all of our roads by surfacing with hard material, but such an extensive improvement is unnecessary, because of the fact that repeated investigations have shown that 20 per cent. of the roads carry 90 per cent. of the traffic.

From the standpoint of the farmer, the increased loads which his team can draw, the possibility of making a greater number of trips per day, and the decreased wear and tear on his team, his equipment and himself should furnish powerful arguments in favour of road improvement.

Practical Value of Road Improvement.—In determining upon the location of proposed improvements a careful traffic census should be taken, so that the most heavily travelled roads may receive the most thorough improvement, and the other roads be improved according to their importance.

Increase in the value of lands adjacent to the public road invariably follows a marked improvement in the road. This increase is unquestionably genuine and not, as many persons claim, a fictitious increase arbitrarily assumed by the assessor, and which imposes upon them an unwarranted increase in taxation. If the increase were fictitious, the farm would have no greater market value than it had before. As a matter of fact the farm, if it were put upon the market, would command a better price than if the improvement had not been made. The very fact that market and shipping points are made more accessible to the farm makes the latter more valuable to the prospective purchaser. This point should need no argument to support it, and rests upon the same reasons which make a lot on an active business street in a city more valuable than a lot of equal size on a little-frequented side-street. The fact that more land can be cultivated, that more profitable crops can be grown, that regular delivery of such perishable products as milk and cream, small fruits, truck products, etc., is made possible adds materially to the value of the land. If the owner sells after the improvement he reaps the benefit of the increased valuation. If he retains the land and cultivates it under this improved condition, his yield in income is greatly increased, to say nothing of his comfort and happiness. Examples are numerous of farm products that have gone to waste because the expense of transporting them to market was greater than the amount which would be derived from their sale. Census statistics show that vegetables yield a return per acre about six times as great as the cereals, while small fruits yield a return over eleven times as great as the cereals. Neither of these two products can be grown to advantage except near a good road over which they can be delivered regularly, quickly, and in good condition to the consumer.

That the agricultural regions which are afflicted with bad roads are not utilising their resources as they should has been ascertained in numerous investigations. A striking example of this was found in an agricultural county situated within easy reach of the cities of Washington, Baltimore and Richmond. The roads of the county were almost impassable at certain seasons of the year, and as a consequence most of the agricultural land was untilled. An inspection of the records of the local railway station at the principal town in the county revealed the fact that the incoming shipments of farm products such as could be produced within the county, exceeded the outgoing shipments by nearly 5,000 tons. In other words, the people of that particular county were actually buying from outsiders the food products which they should be producing and selling.

A factor which should be considered in dealing with the subject of road improvement is the effect of such improvement upon population and the labour supply. The last census figures show that over 46 per cent. of our population live in cities of 2,500 inhabitants or more. The boys are leaving the farm for the more attractive surroundings of the city. The immigrant, instead of settling in the country and thereby affording an adequate labour supply, is staying in the city, and by this unhealthy crowding is lowering the standard of living and of citizenship, and increasing the cost of living by increasing the ratio between the producer and the consumer of food. The rural sections which are improving their roads are not losing in population as are other sections. In an inspection of the returns from the census of 1900 it was found that in 25 counties selected at random showing an average of only 1½ per cent. of improved roads, an actual decrease of 3,112 persons to the county occurred between the years 1890 and 1900; while in 25 other counties having 40 per cent. of their roads improved, located in the same States, an increase of population took place in the same period averaging 31,095 for each county.

Better roads mean better schools, because the attendance is greater and the possibility for fewer buildings and more graded schools is increased. This point is manifested by investigations made by the Government which show that in 5 States having a small percentage of improved roads, 59 out of each 100 of the pupils enrolled regularly attended the schools; while in

5 States having a very high percentage of improved roads, the attendance was 78 out of each 100 enrolled. In some prosperous communities having good roads the little one-room schoolhouses have been supplanted by six- and eight-room, graded schools, and a sufficient amount of money saved to provide conveyances for taking the children to and from school.

The benefits of road improvement are incapable of exact enumeration and definition, but they directly or indirectly affect the life of the rural dweller in every way. If he goes to church the condition of the road has its effect. If illness occurs in his family, the effect of the road is the relative quickness with which medical aid can be secured, and, in many cases, this also affects the cost of medical attention. The social intercourse with neighbours, and the pleasure of driving or automobiling depend upon the condition of the roads. All of these considerations must be taken into account when the question of road improvement is to be decided, and they must be weighed against the burdens incident to the improvement to be made.

CHAPTER III

LOCATIONS, SURVEYS, PLANS, SPECIFICATIONS

Location.—A road should be so located as to permit the passage of traffic from one given point to another with the least possible expenditure of time and energy, but due consideration must be given to the initial outlay in the construction, and the subsequent outlay in the maintenance of the road, so that the total cost will not be greater than the resultant benefits. It must be apparent that many factors enter into the problem, frequently making it difficult for even the most skilled and thorough engineers to determine the right course to pursue.

The economic considerations involved in road location are of two kinds: First, those relating to the accommodation of traffic; second, those relating to the road itself. The first deals with the utility of the road to the community, while the second deals with the cost of construction and maintenance of the road. In the consideration of the traffic requirements, due weight should be given to the relative populations dwelling along possible lines of location, the possibilities of development, agricultural and otherwise, following the location of the road, the necessity of shortening the distance between given points and, lastly, the considerations of pleasure and recreation. The second consideration deals with the relative difference in cost among the various possible routes both for construction and maintenance, and involves not only the question of grades, and the availability of materials, but also the type of construction necessary.

In general the most economic location of a road is that over which the annual cost of transportation, the annual cost of maintenance and the interest on the first cost of construction, together with the annual sinking fund, are lowest. Thus, it will be seen that the problem of road location is one dealing largely with financial considerations which must be given precedence over considerations of an engineering character.

It is ordinarily held that the following principles should be observed in deciding on the final location of a road:

1. Follow the route having the easiest grades.
2. Select the shortest and most direct route commensurate with easy grades.
3. Avoid all unnecessary ascents and descents.
4. Cross ridges in lowest passes.
5. Cross over or under railroads: a grade crossing means danger to every user of the road.
6. Cross streams at most favourable sites, and as nearly at right angles as possible.
7. Carry the balancing of cuts and fills only so far as it will reduce the cost of the total earth work to a minimum. When more earth is needed for a fill, it can readily be secured by slightly widening the cut, and where the cuts are in excess, convenient wastage can readily be found by widening the nearest fill.
8. Do not overestimate the advantage of straightness. The curved road around a hill

is often no longer than the straight road over it. In addition, a more or less sinuous course is an advantage from a maintenance standpoint, as on a winding road the wheel traffic has a tendency to spread over the entire surface, which is seldom the case on a straight road, particularly when the crown is high.

9. Under modern conditions of traffic, sharp curves are a source of constant danger. The radii of curves should never be less than 100 feet, and as low as this only where an unobstructed view can be had of the road ahead.

10. Carry the road along the southern or western slope of ridges, if possible, so that it may be least exposed to storms and dry out more quickly after heavy rains and the melting of snow.

Surveys.—The purpose of a survey is to secure the necessary data for determining the best location, as defined above, to supply such other data as may be needed in the preparation of the plans and estimates of cost, and finally, together with the specifications, to serve as a guide in the actual construction of the road. The cost of the survey will vary greatly with conditions. In the construction of an improved road through a new territory where the route is not clearly defined by natural topography, several surveys may be necessary, while if it involves the improvement of an already existing road, the location of which can not be altered except to a very limited extent, a single survey may be sufficient. The more extensive surveys are usually divided into three parts: the reconnaissance, the preliminary, and the final survey.

The reconnaissance is a more or less rapid examination of the region to be traversed, for the purpose of obtaining information as to the general feasibility of the proposed route, and to secure the data necessary for the rapid and intelligent prosecution of more detailed surveys, should they prove necessary or advisable. Reconnaissance should, in general, include the examination of an area rather than of one or more routes. This is especially true where the road is of any considerable length, for, having familiarised himself with the entire area, the engineer will find no difficulty in choosing the one or more lines for which more detailed surveys are needed in order to determine the final location. For this work the topographical sheets of the United States Geological Survey are extremely valuable, and if the region in question is covered by such a survey, a copy should, by all means, be secured. The topographical sheets, covering approximately an area 30 miles square, can be obtained from the Superintendent of Documents, for the sum of 5 cents each. Sometimes a study of the topographic map will make it possible to dispense with the field reconnaissance entirely, or at least reduce it to a minimum.

In making the reconnaissance the following data should be carefully noted and recorded in the field book: The location and approximate elevation of all low passes; the general trend of all ridges and streams; the inclination of the rock strata; conditions as to dryness, etc. Advantageous bridge sites should be determined; all sources of supply of road material, stone for concrete, water supply, etc., should be carefully noted.

The reconnaissance should determine on one or more lines to be surveyed in detail in order to establish finally the best and most economical route. For these lines an instrumental survey is necessary, or at least advisable. This survey should be accurate enough to mark the exact location of the proposed improvement on the ground, and also to obtain all necessary data for plotting the map and preparing profiles, estimates of the earth work, etc.

The survey usually consists of a transit line with levels and cross-sections taken at every station and at such intermediate points as may be necessary to give the required accuracy in computing the earth work. Full notes are also taken in regard to the width and character of all streams crossed, low and high water marks, all crossroads, private ways, the character of the soil and of any material suitable for road metal or use in constructing culverts or bridges which may be found in the neighbourhood.

Plans.—The surveys should furnish all data for supplying drawings from which the estimates can be closely computed. The necessary drawings consist of a map or plan of the road and as much of the contiguous territory as may be desirable, a profile and a number of cross-sections. If bridges, culverts or retaining walls are necessary, fully detailed drawings must also be made for these structures.

The survey-notes should be so complete that the map, cross-sections, and profile can be plotted rapidly and with sufficient accuracy. It is the poorest kind of policy to depend on the memory to supply lacking data. Everything should be taken in full in the field and entered in the notebook.

The completed plans should be clear, concise and full of information. The profile especially should be a veritable encyclopædia of information, both for the engineer and the contractor. It should show the present ground line, the finished grade, the depths of cuts and fills, the points of change of grade, location of all crossings and watercourses, together with elevation of high and low water levels, etc.

The scale to which the drawings should be made will depend largely on the amount of detail to be shown. For most general purposes, a scale of 100 feet to the inch for the map and of 100 feet to the inch horizontally and 40 feet to the inch vertically for the profile will be found convenient. Where much detail is to be shown, or on very difficult sections, this scale may be enlarged to any desired extent.

The layman is apt to belittle the value of the preliminary work done on the surveys, and in making of plans, etc.; yet these are of the utmost importance and are absolutely necessary for an economical solution of the questions involved. A few extra days spent in this preliminary study of conditions will often result in the saving of large sums of money, not only in the actual construction and in the maintenance of the road itself after it has been built, but in securing a much better route than one which might be secured without such study. Many of the questions involved in highway location are of an extremely difficult nature to solve, and it is needless to say that hard problems cannot, as a rule, be correctly solved without the requisite time and study.

The following instructions for making road surveys are used by the Office of Public Roads, United States Department of Agriculture:

INSTRUCTIONS FOR MAKING ROAD SURVEYS

All surveys for roads which it is proposed to improve with the cooperation of this Office should be made strictly in accordance with the following rules:

All notes should indicate the date on which each part of the survey is made, the names of the men performing the work and the weather conditions. All of the work should be plotted and accompanied with a complete copy of the notes.

LOCATIONS, SPECIFICATIONS 71

TRANSIT AND LOCATION SURVEY.

1. The transit line should be established following approximately the center of the road. At every hundred feet on this line temporary points are to be established. A spike driven into the road through a piece of red cloth or tape is a station mark that can be easily found after several weeks. The measurement of this line is to be made either with a steel chain or tape, with a degree of accuracy of 1 in 3,000.
2. Wherever it is necessary to make a bend in the transit line, the transit instrument is to be set up at the bend, and the angle of the course ahead with that of the rear course measured, always measuring from the back sight around to the right. The angles are to be measured to the nearest minute, and where local disturbances do not preclude doing so, magnetic bearings of each course should be observed.
3. Opposite the points established in the road, and on the side far enough removed to be clear of all construction work, stakes are to be driven. These stakes should be about 24 inches long, and driven for a depth of 12 to 15 inches. The stakes are to be numbered, beginning with zero, each hundred feet to be a unit. The offset distance of centre of stake from the station point on the transit line is to be measured and recorded in the notes to the nearest 0.10 foot.
4. At all bends stakes should be set on both sides

of the road in a line through the point of deflection and at right angles with the back course. These stakes will be used as reference stakes and should have a small nail driven in the top from which measurements to the nearest 0.01 foot are to be taken to the deflection point in the transit line. Reference stakes should be driven flush with the ground and another stake driven near by for a marker.

5. As a rule, deflection points should be made at even stations or half stations, a half station being designated by the number of the previous station with + 50.

6. After the location of the transit line as described, offset measurements are to be taken at each station or as much oftener as may be necessary to locate properly the sides of the travelled way and fences or walls alongside the road wherever such exist.

7. Measurements should be taken so as to locate all bridges, culverts and cross drains of whatever description, and the direction of flow through them should be shown by an arrow. The clear opening of all waterways should also be indicated.

8. The location of all crossroads and private entrances should be indicated.

9. Landowners' names should be obtained and dividing fences, where such exist, should be located.

LOCATIONS, SPECIFICATIONS

LEVELS.

After the transit and location survey is made the levels are run as follows:

10. Permanent bench marks at either end of the work and at convenient intermediate points are to be established well out of the way of any construction. The number of bench marks should be at least four or five to the mile and as much oftener as convenience may require. Bench marks should be on permanent objects on which a rod can be conveniently held, and located where they can be readily identified on the ground. The roots of trees with low-hanging limbs are not convenient, nor is a point so far back from a line of trees along a road as to shut off all view of the bench mark, except directly opposite it.

11. A line of check levels should be run touching every bench mark, and separate notes kept of these check levels. Elevations should check to 0.10 foot per mile. All readings on bench marks and turning points should be to nearest 0.01 foot.

12. Readings for ground elevations should be to nearest 0.10 foot. Ground elevations are to be taken at the centre of the road at each station or 100 feet and as much oftener as may be necessary to show irregularities in the profile or cross-section.

13. At each place where a centre reading is taken

74 ROADS, PATHS AND BRIDGES

side readings are to be taken to show accurately the cross-section of the road.

14. To take a cross-section, first take reading of the rod on the top of the stake at that particular station and a ground reading at same point. Enough readings are to be taken at other points across the line of the road to show the true shape of the banks, gutters and ditches on each side and the road between. The distance of each reading from the transit line is to be recorded as well as the reading itself.

15. Elevations are to be taken of the following points:

    a. The bottom of openings at each end of all culverts, indicating them as east and west or north and south ends.

    b. Bridge floors, tops of abutments and bridge seats.

    c. The entrance and exit ditches on stream bottoms about 25 feet from either end of a culvert or bridge, so as to give the grade of the stream bed near the culvert.

    d. High and low water in streams (estimated).

    e. Water surface of streams as found.

PLAN .... Scale—$1'' = 40'$
PROFILE .... Vertical Scale—$1'' = 4'$
              Horizontal Scale—$1'' = 40'$
CROSS-SECTIONS .... Scale—$1'' = 4'$

Note.—A profile of road with a grade of more than 4 per cent. should be plotted with a vertical scale of $1'' = 8'$.

Specifications.—The purpose of a set of specifications is to set forth in clear and unmistakable language the work to be done, the manner of doing it, and the character of the materials to be furnished. Usually the specifications, together with the engineer's estimate, form the basis on which the contractors bid, and after the contract has been let, the specifications serve as a guide for both contractor and engineer in the further prosecution of the work. In how much detail the various operations are to be specified will depend upon conditions. Sometimes it may be advisable simply to set forth certain standards to which the finished work shall conform. In this event, the manner of carrying on the work is left entirely with the contractor. When there are unknown or hazardous conditions, such as are sometimes met with in the construction of foundations for bridges, or erecting bridges over streams subject to violent floods which may endanger the work in progress, it may at times be considered preferable to let the contractor assume the risk. In this case, however, great care should be taken to secure a contractor of known integrity and responsibility.

More often, however, not only the standard to which the finished work is to comply, but also the character of the materials which are to enter into the construction, as well as the manner in which the work is to be carried on, are prescribed. Oftentimes the specifications are made so complete as to form a perfect formula for the contractor to follow. It should be kept in mind, however, that under such conditions, if for any reason the finished work does not comply with the requirements, the contractor cannot be held legally responsible so long as he has substantially complied with the various specifications. The courts have ruled that a man cannot be held responsible for the results of his work when he is not given any choice in the manner in which it must be done, but must follow regulations in every detail.

The specifications when drawn should be examined, first as a whole, and then each clause separately. No conflicts or ambiguities must exist, and nothing should be inserted which is not necessary. It is a good rule to specify only what is really wanted, and to write these specifications so clearly that there can be no mistake as to what is desired. One clause which is nearly always found in all specifications for road building involving any considerable amount of excavation, and which has in the past caused more friction than almost any other single clause, is that pertaining to the classification of the earth work. Quite often three classifications are given, as earth, loose rock, and solid rock, and sometimes a fourth is added—that of hard pan. These classifications are in themselves all well and good, but the difficulty comes in describing the different classifications in such a manner that in the field one class may be readily distinguished from the next. Because of this difficulty it would seem advisable in road work to limit the classification of earth work as much as possible. The drawing up of proper specifications is no small part of the work of the engineer, and the manner in which they are drawn will often not only save endless friction and hard feelings during the progress of the work, but save much money to the community. They may seem simple at first glance, but the writing of proper specifications requires knowledge, skill, experience and ability.

CHAPTER IV

THE EARTH ROAD

According to a careful mileage census made by the U. S. Office of Public Roads, there were in 1909 about 2,210,000 miles of road in the United States, of which upwards of 2,000,000 miles may be classed as earth roads. It is evident from this showing that the task of surfacing all of our roads with hard material, or even the major portion of them, is so great as to be impossible of accomplishment for a great many years to come. The best that we can do is to classify the roads so that only those which carry the heaviest traffic will be surfaced with hard material, while the remainder will be given such simple and efficient treatment as to render them capable of meeting requirements at small outlay.

Already an enormous traffic is carried over the country roads in the United States, estimated at not less than 250,000,000 tons annually. Impressive as this tonnage appears, it is but a fraction of the traffic which our country roads would be called upon to sustain if they were in fairly good condition. Proof of this is found in the experience of France, where an official census has brought out the fact that the public roads carry one and one-third times as much freight as the railroads. According to the reports of the Interstate Commerce Commission, the railroads of this country carry upwards of 900,000,000 tons annually. If our public roads were used to the same extent as the French roads, it would mean a traffic of about 1,200,000,000 tons annually instead of the 250,000,000 tons, as at present.

An earth road may be defined as a road composed of natural soil, to which no other kind of surfacing material has been applied, and with which no binder or filler has been mixed. It differs from a sand-clay road to the extent that the latter is composed of sand and clay mixed in suitable proportions.

Location.—It is important that the road be located so as to serve the needs of traffic best, to permit due economy in construction and maintenance, to obtain a grade as nearly level as practicable, and to permit thorough drainage. The considerations which should govern

THE EARTH ROAD

the location of roads are fully dealt with in Chapter III. By far the greater proportion of our roads have been located at haphazard, in many cases following Indian trails, paths of wild animals and farm boundaries. This is particularly true in the Eastern States, which were settled first. The result is that instead

Typical Section for Side-Hill and Gradual-Slope Locations.

of the roads being adapted to the traffic requirements, the traffic is compelled to adapt itself to the road. In the West the roads are laid out on section lines. These sections are all square, and consequently the roads are all at right angles. If a person desires to cross the country, it is necessary for him to follow the boundaries of a series of rectangles, instead of going directly to the point he desires to reach. If it were possible to relocate the public roads according to the needs of traffic and agricultural development, so that distances might be shortened and easier grades obtained, our total mileage could be cut down at least 100,000 miles, and communication rendered far less difficult and costly in every section of the country. Such general relocation is impossible, and the best that can be done is to avoid unwise location of new roads, and to relocate old roads whenever conditions require and permit.

Grades.—The term "grade," as used in this chapter, means the slope of the road along its length. A steep road would be described as a road having a steep grade. Among road builders the grade is expressed in terms of percentage, as 1 per cent., 5 per cent., 10 per cent.; each per cent. meaning a rise of 1 foot in each 100 feet of length.

In the construction of new roads and the regrading of old roads it is customary to specify a certain per cent. as the maximum grade allowable at any point on the road. A minimum grade means the least that can be allowed for good drainage. A number of considerations are influential in determining the maximum

THE EARTH ROAD

grade which may be allowed, but the most important is expense, as it is necessary to adapt the work in hand to the means available. The topography of the country has an important bearing on the question of grades, as a much more nearly level road can be specified in a flat or rolling country than in a mountainous region. The character of the soil also has some bearing upon the question. By common consent it is agreed among highway engineers that no road should exceed a grade of 5 per cent. except in extreme cases where, by reason of natural difficulties or lack of funds, it is impracticable to reduce the grades to that point.

Steep grades are a powerful handicap to traffic, and wherever possible they should be eliminated. While it is a matter of common observation that the load which a team of horses can draw on a steep hill is very much smaller than the usual load on level ground, it is not generally known that on an average macadam road it requires approximately four times as much power to draw a load up a 10 per cent. grade as is required to draw the same load on a level. This means that an eight-horse team would be required to draw up a 10 per cent. hill the same load that two horses could draw on a level road. It may be said, in modification of this, however, that for short distances a horse is able to exert about twice his natural pull, so that if the grade were short, four horses might, by exerting their maximum pull, accomplish the same result. But the loss of tractive power on steep grades is greater than shown by theory, since the power of a horse decreases very rapidly on steep inclines. The leading authorities on highway engineering express the matter about as follows: Assuming 1,400 lbs. to be the load which one horse can draw on a level earth road, he should be able to draw 650 lbs. on a 5 per cent. grade and 340 lbs. on a 10 per cent. grade, with about the same degree of ease.

While it is frequently expensive to obtain easy grades, the fact should be borne in mind that the work is permanent in character. No matter what surface material may be applied to a road, it will wear out and have to be replaced. Not so with the grade; it is a permanent step, not only toward the building of a good earth road, but also toward any type of improved road which may be determined upon at some later time. A steep road is much more difficult to maintain than a road with a flatter slope, as the former is much more likely to be damaged by the action of water, which tends to wash and gully the surface. The injurious action of horses' hoofs and narrow-tired, heavily loaded wagons is also more pronounced on steep grades.

There are three ways by which an easy grade may be obtained: First, to locate the road so that it will go round the hill, instead of over it; second, to have it run diagonally up the face of the hill, doubling back and forth a sufficient number of times to keep the grade down to the desired per cent.; third, to cut down the hill.

Another plan, which might be considered a modification of the second, is to begin the ascent of the hill quite a distance from the base. It is a matter of common observation that many country roads run straight to the base of a hill before beginning the ascent. In almost every case they could, by leaving a straight line some distance back, approach the hill on an easy grade. The question of cost will largely determine which of these three methods should be adopted. If the hill is a long one, it will usually be found cheaper and more practicable to go around it. This will not necessarily result in lengthening the road, as shown by the familiar example of the bucket bail, which is the same length when resting on the rim of the bucket as when in a vertical position. If the hill is short, it will probably be cheaper and more satisfactory to cut it down, using the material from the cut to fill in the approaches on each side. Where the road leads from lower ground to a plateau, the method of carrying a road up the face of the hill diagonally will sometimes be found most feasible, but each case must be decided in conformity with the local topography.

If it is necessary to lengthen the road to even a considerable extent in order to secure easy grades, it may be found in many cases to be real economy to do so. The same energy which would be expended by a horse in drawing the load up a steep grade would suffice to draw it a far greater distance on a comparatively level road. Many scientific tests have been made to demonstrate this in exact terms. The point will be made sufficiently clear, however, by the statement that to lift a ton a distance of one foot requires an expenditure of energy amount-

THE EARTH ROAD

ing to 2,000 foot-pounds. Therefore, in drawing a ton a distance of 100 feet on a 10% grade, the load would have to be lifted ten feet, involving an expenditure of 20,000 foot-pounds of energy, and all this is in addition to the force required to draw the load a distance of 100 feet on a level. It must, therefore, appear that the burden imposed by distance is not nearly as great as that imposed by steep grades. Of course it should be borne in mind that a material lengthening of the road may add to the cost of construction and the cost of maintenance. The best course is to give due weight to all factors in the problem.

Drainage.—Water is destructive to all roads, and particularly to earth roads, so much so that good drainage is the keynote of success in road construction. To remove quickly the water which reaches the surface of the road, and to intercept the flow of water from higher grounds toward the road, a system of surface drainage must be provided. Water attacks the foundation of the road as well as the surface, in many cases, and to meet this danger subdrainage must be provided. The subject of drainage is, therefore, subdivided into surface drainage and subdrainage.

Surface Drainage.—Most country roads are too flat to shed water; in fact, many of them are concave, owing to the fact that traffic is kept consistently in the centre and wears down the surface until the road is more in the nature of a ditch than a highway. As the roads are usually repaired only once or twice a year, grass and weeds are permitted to grow close up to the travelled way, still further preventing the flow of water from the road to the ditches.

If the road is comparatively level, so that the water stands upon it, the surface soon becomes soft, causing deep holes and ruts to form under the impact of traffic. When this incipient damage is done, every heavy rain thereafter hastens the destruction of the road, because the water follows the wheel ruts, widening and deepening them. Eventually, if preventive measures are not taken, this will totally destroy the road. In any event, under such conditions the cost of repair will be large.

This damage to the surface can be easily prevented by giving the road a crown or slope from the centre to the sides sufficient to cause the water to drain quickly to the side ditches, instead of running down the middle of the road; but it is necessary to exercise judgment in determining upon the slope or crown to be adopted. If the crown is made too steep, the water will rush off to the side so quickly as to cause damage to the shoulders or sides of the road. If it is too slight, the water will flow down the centre instead of to the sides. In a perfectly flat country a somewhat slighter crown is necessary than on the hillsides, because in the former case there is no tendency of the water to flow down the centre, while in the latter case the slope at the sides must be at least equal to the longitudinal slope. Otherwise, the water will follow a diagonal course and may carry off some of the surface material. The best practice is to allow a slope, averaging from ¾ inch to 1 inch to the foot, but the individual judgment is necessary to determine whether it is advisable in specific cases to increase or decrease these standards slightly. The road builder should avoid the mistake of crowning his roads too steeply, not only because of the consequent damage to the shoulders, already referred to, but because in such cases the wagons will "track" or keep to the centre and eventually cause the road to be flat or hollowed out on the most heavily travelled portion.

A natural mistake is sometimes made by reason of the literal following of text-books in giving a *uniform* slope from the centre to the sides, which results in making the road like a roof, in which the centre of the road forms the comb. In actual practice the road should be curved and the total slope from centre to sides should be such as to give the required *average* slope per foot. By actual measurement it might appear that the slope will be only ⅛ of an inch to the foot near the centre and considerably more than an inch to the foot at the side of the road. This is all right, as long as the road maintains its convex shape. On sharp hillside curves it is usually advisable to give a single inward slope to the road.

Ditches.—The next most important point in providing for surface drainage is to construct suitable side ditches. All these side ditches should have a fall or slope of at least six inches

THE EARTH ROAD

in each 100 feet in length; if the fall is less, the water will not flow quickly enough and trouble will be had, particularly in winter or early in the spring when the snow melts. These side ditches must be ample in size to provide for the greatest volume of water that may reasonably be expected by reason of heavy rains, storms or the melting of snows. In order to provide sufficient capacity, the ditches should be made wide, rather than deep, as deep ditches beside a road are dangerous to traffic and are more expensive to construct and maintain. The best plan is to have frequent outlets from the ditches, either by means of culverts, pipes, or by turning the ditches into lower ground, rather than to allow the water to flow along the road any great distance.

Five 12-inch pipes in a mile of roadway are about as cheap and far more effective than one 24-inch pipe, because the water is disposed of before it gains force or headway or has time to damage the road. The maximum velocity for a 24-inch vitrified tile, flowing full without head on a grade of 1-inch per 100 feet, is 3.6 feet per second, or about 2½ miles per hour; when the grade or slope is increased to 36 inches in a distance of 100 feet, the velocity becomes 20 feet per second, or about 13¾ miles per hour. The discharge for the 24-inch pipe in the first instance, will be 5,086 gallons per minute, while in the second instance it will be 28,260 gallons per minute. It will, therefore, be seen that a 24-inch pipe, laid on a grade of 36 inches to the 100 feet, will have over five times the capacity of the same pipe laid on a grade of 1 inch to the 100 feet.

Under the same conditions, the maximum velocity for a 12-inch tile on a grade of 1 inch per 100 feet, equals 1¼ feet per second, or about ⅞ of a mile per hour, and for the same tile on a grade of 36 inches to the 100 feet, the velocity would be 7½ feet per second, or about 5⅛ miles per hour. The discharge for the 12-inch tile in the first instance would be 442 gallons per minute, and in the second instance 2,650 gallons per minute, or about five times as much. It will thus be seen that comparing the 12-inch pipe and the 24-inch pipe on a grade of 36 inches to the 100 feet, the five 12-inch pipes would remove in the aggregate 13,250 gallons per minute, as compared with 28,260 gallons per minute by the one 24-inch pipe, but the advantage of the former lies in the fact that the water is removed at five points instead of one.

Another important point in the foregoing is, that by increasing the grade or slope of the pipe, the capacity for removing the surface water is enormously increased. In order to protect culverts or pipes from damage, when discharging water under full pressure, or when a culvert or pipe is given a considerable slope or grade, it is desirable that the joints be cemented, if a pipe is used, and that the ends of the culverts be protected with masonry, or concrete wing walls. In addition to this the spillway should be paved with cobblestones, in order to prevent washing. Another point in favour of having a sufficient fall or slope to the culverts is that they will be self-cleansing and so keep open. A culvert laid flat may soon fill up.

Mud-holes cannot be successfully drained, as a rule, with culvert pipes. The best plan is to throw out the soft mud and replace it with good firm earth, so that it becomes level after consolidation with the surrounding surface. The ditches should then be sufficient to drain the mud-holes and carry the water to the culverts.

The sides of these ditches should have an easy slope, particularly on the side of the road, as this will tend to prevent accidents to traffic, as well as the caving in of the banks of the ditches. The construction of deep ditches should be avoided. In most cases ditches made with the road machine in shaping up the surface of the road is sufficient for surface drainage.

Culverts should be built at the low points where outlets are available and existing streams should always be utilised for outlets, when possible. Where only a slight volume of water is to be removed from ditches, it may be carried under the road in tile pipes instead of concrete culverts. In such cases the pipes should be laid deep enough to prevent their being broken by the traffic. If it is impossible to place the pipes deep enough to be removed from the effect of traffic, it is well to use concrete. The construction of small culverts and drains will be more fully explained in a separate chapter.

Subdrainage.—Many thousand miles of public road in the United States are located on low, swampy ground, or on ground which possesses very poor natural drainage. A large percentage of prairie roads are in bad condition for several months each year, by reason of a wet subsoil. In such cases surface drainage, no matter how effective, will not always serve to keep the road in good condition. Consequently, it becomes necessary at times to install

THE EARTH ROAD

a system of underdrainage, so as to clear the soil of surplus water and to give the road a solid, dry foundation. Water in the subsoil becomes ice in winter and expands, thereby heaving the road. In the spring the ice melts and the foundation becomes softened to such an extent that the whole road gives way, or, as it is generally stated, "the bottom drops out of the road." It can readily be seen that the maintenance of roads under these conditions is exceedingly difficult and very expensive. Underdrains can usually be provided at small expense and will last quite a long time, if properly maintained. Many roads are greatly damaged by springs in the soil. These should be tapped by blind drains of stone or pipe, and the water carried diagonally to the side ditches.

Hillside roads are often subjected to the destructive action of water, which drains from the higher ground into the foundation of the road. Surface drainage is, in some cases, sufficient to protect the road. Where it is not sufficient, the best plan is to dig a deep ditch some distance above the road on the hillside, of sufficient capacity to intercept and carry off the flow of surface water. This ditch should be given outlets to lower ground at frequent intervals.

The usual method of underdraining a road is to provide a narrow trench on each side at the bottom of which a pipe of from 4 to 6 inches in diameter is placed. Ordinarily a 4-inch pipe will be found sufficient. These pipes are usually composed of terra cotta tile. The depth to which these pipes should be laid depends largely upon the character of the soil and the depth of the frost line, but in general it should be about 3 or 4 feet. The pipe should be laid near the bottom and the trench then filled with broken stone, gravel or broken brickbats. The pipe should have a fall of not less than 6 inches for each 100 feet of length. It is unwise to give too much fall to small drain pipes, as the swift current may wash away the ground about the drains and displace them. The sides of the trenches should slope gradually, as this will prevent the ground from caving in and will also give greater stability to the drain pipes. The outlets or spills for the pipes should be paved so as to prevent washing.

Care should be taken to lay the tile true to grade as, otherwise, it will soon become ineffective. Wherever the pipe sags it will soon be filled with sediment, and if there is a crest the silt will accumulate immediately behind it. The ends of the pipes should be covered with an iron grating, which will prevent vermin from entering.

Where pipe drains or concrete culverts can not be provided, it is sometimes practicable to construct blind drains with flat stones.

Some authorities recommend a line of tile under the centre of the road, but this is not practicable as a rule, as it is much more expensive, and involves a greater amount of digging, both in the original installation of the pipe and in repairs, should they become necessary. Furthermore, if the road is later on surfaced with hard material, it will become increasingly expensive to reach the pipe for repairs.

A method which may be practised to advantage in soil which is free from rock and easily worked, and where the ground is practically level is to grade the road up in the form of an embankment above the level of the surrounding country. The water may then drain from the road instead of to it.

Width of Road.—The width of right of way is specified in most of the States by statute, but is usually not less than 40 and not more than 66 feet. The width of the travelled roadway is much less, as allowance of at least six feet on the outside of each ditch should be made for footways. It is advisable to have the road wide enough to meet all traffic requirement, but it is a mistake to have the travelled way exceptionally wide, as this will necessitate deeper ditches, and will not only be more costly to construct but also to maintain.

Clearing Roadway.—After determining the width of roadway, ditches and footways, the next step is to remove all stumps, brush, roots, rocks, etc.

Cuts and Fills.—The practical road builder always endeavours to establish the grade of the road so as to make the cuts and fills equal, as otherwise waste of material will result. If the cuts are greater than the fills, the result will be a greater amount of loose earth than can be used, while if the fills are greater, it will be nec-

THE EARTH ROAD

essary to obtain additional earth from borrow-pits and haul it to the road. Some of the cuts and fills are so far apart that it is cheaper to obtain material from the side of the road, or to waste the material from the cuts, rather than endeavour to balance the cuts and fills.

An important point to be considered in connexion with excavation is that loose earth at first occupies greater space than compact earth, while the final result is that there is considerable shrinkage in fills. It is a curious fact that the earth in a fill or embankment actually compacts to less space than it occupied in its original position. The amount of shrinkage varies with the character of the soil and is about as follows: Gravel or sand, 8 per cent.; clay, 10 per cent.; loam, 12 per cent.; loose surface soil, 15 per cent.; and puddled clay, 25 per cent.

The side slopes on cuts and fills must be given an angle which will insure stability and at the same time cause the least waste of material. The method of determining the angle of the slope is by the ratio of the horizontal distance to the vertical distance. We can assume the side of the embankment to form the hypothenuse of a right-angled triangle. If the vertical line of the triangle is 1 foot and the base line extending out to the point of the slope is 1½ feet, the slope should be designated as 1½ to 1. The slope will, of course, vary with the nature of the soil. Common earth will stand a slope of 1 to 1, but it is safer to make it 1½ to 1. Gravel requires a slope of 1½ to 1, while clays vary widely, ranging from 1 to 1 to a slope as flat as 6 to 1. In general practice the slope of 1½ to 1 is found best. It is well to sow grass seed on slopes, or, if that is not practicable, to sod them, as greater stability will be obtained in this way.

For slight cuts and fills, where the soil will permit, it will be found that the road machine or road grader is sufficient; where the soil is hard or mixed with pebbles or field stone, it is frequently found to be economy to use a road plough and follow it with the road grader. For excavation on a larger grade the slip scraper is exceedingly useful, and where the cuts and fills are considerably apart the wheel scraper will be found most useful. The wheelbarrow is rarely used, except for small jobs or in wet, swampy places. The elevating grader is frequently used to advantage in prairie regions and in low, flat ground free from rocks, as it elevates the road above the surrounding country, and thereby promotes good drainage. It can also be used to advantage in load-

THE EARTH ROAD

ing wagons on cut-and-fill work. Road-building equipment and its use are taken up in detail in a separate section of this chapter.

When the roadway has been cleared and brought to a desired grade, careful examination should be made of the surface at all points, and wherever it is found to be soft, spongy, or insecure, the soft material should be removed and replaced with good, firm earth, sand or gravel. The material should then be tamped in place until the surface is smooth and compact.

Road Implements and Machinery and Their Use.—Road building has been much simplified and cheapened by the substitution of machinery for hand labour in transporting material from place to place. The first advance over the burden bearer was the wheelbarrow. An American engineer in the Philippine Islands tells of his experience with the native labourer when the wheelbarrow was introduced there as part of the regular road-building equipment. As soon as the barrow was loaded two of the natives bravely picked it up, carried it to the point where the material was needed, emptied it and carried it back; the wheel at the end of the barrow had no significance to them. Modern practice has made even greater strides to-day. The wheelbarrow is rarely used now, except for small jobs or in wet and swampy places. It has been superseded by some form of drag scraper drawn by horses.

The complete road-building outfit consists of a great number of units, which may be roughly enumerated as ploughs, drag and wheel scrapers, road graders or road machines, disc harrows, dump carts, elevating graders, sprinklers, rollers, crushers with elevators, screens and bins; and in the construction of roads of modern type with bituminous binders a great deal of special equipment has been devised such as tar and asphalt spraying machines, and tank-wagon.

Scrapers are intended for use in moving material after it has been loosened by ploughing. They are of two kinds—drag scrapers and wheel scrapers. The drag scraper is made in several sizes running from about 3 cubic feet capacity to a capacity of from 5 to 7 cubic feet. The average cost is from $6 to $7 each. The smaller size is designed for one horse and the larger sizes for two horses. The drag scraper is used for moving earth short distances.

The wheel scraper may be described as a steel box on wheels, open in front, and provided with levers by which the box may be raised and lowered and its contents dumped. The capacity is usually from 9 to 16 cubic feet. The cost should be from $25 to $40.

The elevating grader is provided with a frame, resting upon four wheels, from which is suspended a plough and frame carrying a wide travelling belt. The plough loosens the earth and casts it upon the inclined belt, which in turn carries it to the embankment, or to the wagons, as the case may be. This machine is particularly adapted for the construction of earth roads in a prairie country. It cannot be used to advantage in a very hilly or rocky country.

The disc harrow is used mainly in the construction of sand-clay roads for the purpose of thoroughly mixing the sand and clay. Its use will be explained in the chapter on sand-clay roads.

The steam roller, the sprinkler and the

crusher, with its appliances, are mainly used in the construction of gravel and crushed-stone roads. It will be described in greater detail in the appropriate chapters.

For use about the farm and in the treatment of ordinary earth roads, the plough, the drag scraper, the wheel scraper and the road machine are the implements most generally used. A split-log drag is very simple and exceedingly useful in the maintenance of earth roads. It will be described in another chapter.

Under certain conditions the plough is a most useful implement in road work. When the soil on the surface of a road is excessively sandy and the subsoil is of clay, or of gravel and clay, the road will be greatly benefited by a deep ploughing. The ploughing brings the clay from beneath and mixes it with the surface soil and sand. Thus a sand-clay road is formed at small expense. On the other hand, if the road is entirely formed of deep sand, it will prove a very great mistake to plough up the road bed unless clay can be added. The ploughing only deepens the sand and breaks up what little hard surfacing material has been

THE EARTH ROAD

formed on top. Again, if the surface has only a little sand or gravel in it, and the subsoil is practically pure clay, it will prove a great mistake to plough it up. To do so would bring an excess of clay to the surface, and effectually destroy the surface coating of sand or gravel or soil. These are apparently very small and insignificant matters, yet a correct understanding of them, with regard to the principles involved, will enable a road foreman to improve the roads under his charge.

When ploughing is undertaken, the best method is to begin in the middle and "back-furrow" both ways to the middle, thus forming a crown. After the road has been ploughed, it may be harrowed and carefully smoothed. Ploughing should be done in the spring or early summer. A plough can be used in ditches to advantage. In excavating there is no better way to loosen earth than by the use of the plough. For this purpose there are various kinds in use. The old-fashioned coulter plough is effective in breaking up hard gravel or other material.

When it is necessary to make ditches wide

and deep, nothing has yet been devised better than the ordinary drag scraper. It is serviceable in hauls under seventy-five feet long for making fills. Frequently a road becomes worn down and requires widening. The sides may be ploughed and the earth pulled in with the scraper. When both sides have to be pulled in, it is a good plan to make a circular trip, pulling in the earth from both sides at the same time. Ditch work may often be handled in this way and greatly facilitated. Two horses and two men will handle many times more earth than could be handled if work was done by hand with shovels. It is a mistake, however, to attempt to handle material in long hauls with a drag scraper. The wheel scraper is better adapted to such hauling, but still should be limited to about 1,000 feet haul. Furthermore, the wheel scraper is not well adapted to ditch work, for the reason that the wheels require a greater width than is usual in ditches. These scrapers are better adapted to grading and handling earth where many cuts and fills are necessary. As a rule, it does not pay to work less than four to six in a run, because an extra team is necessary to help in loading, and with a less number of scrapers this team is idle much of the time.

The road machine is one of the most generally used of all road implements and scarcely needs any description. It may be briefly said to consist of a frame on four wheels, supporting an adjustable blade, the front of which cuts a furrow while the rear end pushes the earth toward the centre of the road and distributes it. The work of the grader is superior to that of the plough and the drag scraper, as the cut is uniform, whereas the plough cuts irregularly, and moves material in much larger quantities and at less cost than can be done by the drag scrapers. In using the road machine or road grader, it is best to put not more than from 4 to 6 inches of loose earth into the road at one working. The grading should be done early in the summer when the soil is damp. The loose earth will then pack and bake; it will not be so liable to become dusty in summer, and will have ample time to settle before the rains begin in the fall. This is one of the most important points in the whole problem of earth-road construction.

The road machine is on the market in every conceivable design and varying in size from a machine suitable for two horses to one capable of withstanding a traction engine. It is most useful in crowning and smoothing the road and for opening ditches. What was said with reference to the use of the plough is also true in regard to the use of the road machine. It is unwise to pull loose sand upon a sandy road, for to do so is only to make it deeper. On the other hand, if there is clay in the subsoil under the sand, it will improve the road to pull it up with a road machine. It is likewise bad management to pull clay upon a thin coating of gravel or soil. The clay will hold water and make the gravel soften.

It is a great mistake to pull clay from ditches upon a macadam surface with a road machine. For the same reason, it is a mistake to use a road machine indiscriminately and to pull material from the ditches upon a sand-clay or gravel road. Frequently turf, soil and silt from the bottom of the ditch are piled in the middle of the road in a sort of ridge, or, if any effort has been made to spread it, ofttimes it is done in such a way as to make matters worse. Material containing grass or other vegetable matter should never be allowed to be placed on the road, unless it is a sand road and no clay is obtainable. Weeds and grass should be burned or cut and removed before grading is begun. This simple plan will do much to relieve the objection often met with in working the road in the fall, when the ditches are filled with grass. To pull this mass of weeds and grass and sediment into piles on the travelled track, besides making it uneven is the best way possible to start mudholes.

Another important point in building up a road with a road machine is to avoid building up too much at one time. It will be found that a road built up after using the road machine a number of times will stand far better than one built all at once. In the first instance, the material is brought up in thin layers and firmly packed before the next layer is brought up, and in that way the road is made up of a number of thin layers, each one of which is well puddled and packed before the succeeding layer is added. It is too often the case that the road builder thinks he must have his road high in the first instance and, consequently, piles up 10 or 12 inches of raw material at one time. The result is that when the rains come there are no fewer inches of mud in this newly worked road. This would not have occurred had the road foreman taken more time and built up the road by degrees.

It is also a common mistake to crown too high with the road machine. This is particularly noticeable when the road happens to be a little narrow. For this reason a road to be worked with a road machine should be of ample width, not less than 20 feet anywhere, and better from 20 to 24 feet wide.

CHAPTER V

THE SAND-CLAY ROAD

A SAND-CLAY road is composed of sand and clay mixed in such proportions as to form a compact and firm support to traffic. The perfect sand-clay road should be neither sticky nor sandy. The sand and clay may form a natural mixture, in which case, the road is termed a *natural sand-clay road*. The two materials may have become mixed in the fields along the road by successive cultivation of the soil, and if this soil is used in the construction of a road, it is known as a *top-soil road*. There are many varieties of clay, and consequently a wide variation in the characteristics of a sand-clay road. The quality of the sand is a variable factor, as it may range all the way from fine dust-like particles to coarse grains and gravel, and may be perfectly clean, or mixed with loam and other material. In consequence of these wide differences in the materials constituting sand-clay roads, it is impossible to maintain a uniform standard as to quality of the road, or the methods of construction.

Properties of Sand.—Sand is, in general, composed of tiny grains of quartz. While quartz is one of the hardest minerals known, it possesses practically no binding or cementing power. The grains of sand, instead of cohering in a tough mass under the impact of traffic and the action of water, remain loose and shifting. Fine sand, when dry, is easily displaced by the wind, which produces in this way the ever-shifting sand-hills. No road is so difficult to travel as the road located through fine sand, and the difficulties are enormously increased when high winds prevail.

Properties of Clay.—Clay is a decomposition product of the mineral feldspar. If the clay has been carried by running water and deposited as sediment, it is known as "sedimentary." If the feldspathic rock has disintegrated in place, the clay is known as "residual." The sedimentary clay is finer grained than the residual, and is more sticky and plastic. In contrast with sand, which possesses no binding

THE SAND-CLAY ROAD 113

power, but is very hard, clay is a powerful binder. It does not, however, possess the quality of hardness.

It is evident, then, that in the construction of a sand-clay road the important property in the clay is its plasticity or tendency to become sticky and elastic when mixed with water. The clays which are most plastic are called "ball" clays. Another important property possessed by clays in widely varying degrees is the porosity, or capacity for rapid absorption of water. Clays which possess this quality in the highest degree fall to pieces under the action of water, and are called "slaking" clays. It will readily be seen that the plastic or ball clays will form a better and more powerful binder for sand-clay roads than will the slaking clays, but, on the other hand, they will be much more difficult to mix, as they disintegrate with far less rapidity.

The shrinkage of clay is an important characteristic in connection with the building of roads. When water is mixed with clay, expansion results and, when the water evaporates, the clay contracts. This characteristic of expansion is much more pronounced in some clays than in others. It must be apparent that the clays which expand the least are preferable for road building, as they result in the least displacement of grains of sand, and, consequently, tend least to destroy the bond between the sand and clay.

Gumbo or Buckshot Soil.—This is of sedimentary formation and carries a considerable quantity of organic matter. A large area in the valley of the Mississippi river and its tributaries is composed of this kind of soil. The gumbo soil is composed of very fine particles, the colour of which ranges from grey to black, according to the amount of organic matter. Water is readily absorbed and causes the material to become exceedingly sticky; when dry, it breaks up or becomes granulated, which causes it to be termed "buckshot soil."

Sedimentary Loam.—In addition to the gumbo, which contains no sand, there are the sedimentary loams, which include all classes between gumbo and clean sand. As the percentage of sand increases, the characteristics of the buckshot soil are less pronounced. The sand prevents granulation of the soil, as well as marked contraction or expansion. Where the sedimentary loam contains a large proportion of sand, a reasonably good road can be made without the addition of other materials.

Mixing of Sand and Clay.—The theory of the sand-clay road is very similar to that of the macadam road. In the latter rock-dust and screenings fill the voids between the angular fragments of stone and when wet serve as a cement or binder. The grains of sand may be likened to the angular fragments of stone and clay to the rock-dust binder. In the most successful sand-clay road just a sufficient amount of clay is used to fill the voids between the grains of sand. In this way the sand sustains the wear, while the clay serves as a binder. If too much sand is used, the result will be loose sand on the surface; if too much clay is used, the surface of the road will become sticky after rains.

The best mixture of sand and clay can be made when the materials are wet, and particularly is this true of the plastic or ball clays. The more water used the better the mixture and, if practicable, the materials should be puddled.

If the road to be treated is sandy, clay should be hauled upon it, spread as uniformly as possible, and all large lumps should be broken up. As soon as a heavy rain has softened the clay, a few inches of sand should be placed on it and then a thorough mixture should be brought about by means of a plough and a disc harrow. The result will be a successful mixture and a very disagreeable pasty mud. This condition will last for only a short time, and the road will eventually be all the better for it. The extent to which the mixing should be carried on will depend largely upon the character of the clay. If it is a plastic or ball clay, much greater effort will be necessary to obtain a complete mixture; if, on the other hand, it is a slaking clay, the mixture will be much more readily obtained. This kind of clay is not as satisfactory, however, as the ball clay, as its binding powers are much less. In selecting clay for road purposes, it is always best to select the stickiest clay available. A familiar test is to wet the thumb and place it against a piece of clay. If the clay sticks to the thumb, it is reasonable to sup-

THE SAND-CLAY ROAD 117

pose that it will stick to the sand; if it will not stick to the thumb, it is safe to assume that it will be a poor binder in a sand-clay road.

As the desirable proportions of sand and clay are such that the particles of clay barely fill the voids between the grains of sand, it is well, in determining the quantity of clay to be applied to a sand road, or sand to be applied to a clay road, to know approximately how much is needed. A simple method for determining the relative quantity is to take two glasses of the same size and fill one with the dry sand which it is proposed to use, and the other with water. The water should then be poured carefully in the glass of sand, and allowed to trickle down through the sand until it reaches the bottom of the glass. When the water has been poured into the glass of sand to the point of overflowing, we may assume that the voids between the grains of sand have been filled, and, consequently, the amount of water taken from the full glass would represent the volume of clay needed to fill the voids in a volume of sand equal to that in the other glass. It is better to use a little more sand than would appear to be necessary, as the tendency is to underestimate the amount needed. In general practice, clay is placed on a sandy road to a depth ranging from 6 to 10 inches, while sand is placed on a clay road usually to a depth of from 6 to 8 inches.

Construction of a Sand-Clay Road.—The method of construction depends upon whether the subsoil consists of sand or of clay. Good drainage is an essential feature of the sand-clay road, just as it is of all other types of road. A sandy or gravelly soil affords better natural drainage, and if the sand is present to an exceptional extent, the only provision necessary for drainage will be to crown the surface of the road in the same manner as prescribed for earth roads. If the road is located through land that is so low as to be continually wet, it will be necessary, in addition to crowning the road, to provide wide ditches on each side, and to raise the roadbed a little higher than the surrounding ground.

Drainage of a clay subsoil should be provided in exactly the manner as for earth roads in Chapter IV.

THE SAND-CLAY ROAD 119

After proper drainage has been secured, the roadbed should be crowned, beginning near the source of supply of the clay or sand. The clay should then be spread to a depth of from 6 to 8 inches in the centre, sloping off gradually to a thin layer at the sides. Upon the clay should be placed a thin covering of sand. If the clay is of the plastic kind, it will then be necessary to plough and harrow it, taking advantage of rains to puddle the surface with a disc harrow. Sand should be gradually added until the surface of the road ceases to ball and cake. After the road is completed, if it loosens in dry weather, more clay should be added. The mixing of the sand and clay may be left to traffic, but this is an unwise procedure, as it means a very unsatisfactory road for a long period of time.

If the clay is placed on sand to a depth of 6 inches, a cubic yard of clay will cover 54 square feet, consequently, an 18-foot road, treated in this manner, would require 1 cubic yard of clay for each 3 feet of length. A mile of 18-foot road would, therefore, require 1,760 cubic yards of clay. The amount that can be hauled by the average team varies from two-thirds to one cubic yard, according to the character of the road over which the hauling is done.

If the clay subsoil is to be treated with sand, it should be ploughed and harrowed to a depth of about 4 inches. On this prepared subsurface should be placed from 6 to 8 inches of clean sand, spread thickest at the centre and sloping to the sides, in much the same manner as the clay is applied to a sand road. These materials should then be mixed dry instead of wet, which is preferable when clay is applied to sand. Dry mixing is preferable because the clay can be better pulverised when in a dry state. After the dry mixing has been completed, the road should be puddled with a harrow after the first heavy rain. When the materials are thoroughly mixed and puddled, a road machine or grader should be used to give the proper crown to the road. If a horse roller is available, the road can be improved by the use of it. As it is impossible to determine exactly the proportions of sand and clay to be used in the first place, it is necessary to give careful attention to the sand-clay road for a considerable time after it is completed, in order that additional sand or clay may be applied as needed.

Sand-clay roads have been built in the South at costs varying from $200 to $1,200 per mile. This wide variation in cost is due to the difference in the proximity of sand and clay, cost of labour, weather conditions, efficiency of labour, management, etc. Under average conditions a sand-clay road 12 feet in width should cost from $500 to $600 per mile.

The same considerations which should govern in the location of earth roads and in the avoidance of steep grades, apply with equal force to the sand-clay road.

Sand Roads.—Where roads are composed of deep sand, and where clay is not available, it is impossible to make the road satisfactory for traffic, but it is possible to make at least a slight improvement.

Dampness is beneficial to a sand road, and it is well known that wet sand is easier to travel over than dry sand. Consequently, it is better to reverse, to a certain extent, the rules of drainage which apply to earth and sand-clay roads.

The surface of a sandy road should be level and may even be slightly concave, provided the longitudinal grade of the road is very slight. Otherwise, to make a road concave would simply be to transform it into a ditch, which would soon be cut into deep gulleys. Fortunately, in almost all cases, sandy roads are naturally level.

Shade is injurious to roads composed of clay or loam, as it prevents the road from drying out. A sand road, however, should have as much shade as possible in order to prevent it from drying out. In order to overcome the shifting, unstable character of the sand, grass should be encouraged wherever possible. In fact, any vegetable matter that can be made to grow on a sand road, or close up to a sand road, is beneficial. Even if the roots do not spread out into the travel way, the leaves and twigs from bushes will fall into the road and aid to a slight extent in providing a binder. If the road is sufficiently wide, half of it could be planted in grass, and traffic could be required to use the other half; when the grass is mature, traffic could be shifted and the other half planted to grass. Any vegetable fibre on a sand road is beneficial, but is of necessity only a temporary expedient.

CHAPTER VI

THE GRAVEL ROAD

GRAVEL consists of small, partially rounded fragments of stone produced from larger bodies of rock through the action of ice or water.

The best gravel beds are found in the Glacial Drift, which covered Canada and that portion of the United States north of a line running from the Atlantic coast a little south of New York City, in an irregular direction to Cincinnati, thence through Topeka, Kans., and north and west to the Pacific Ocean. The glacial icesheets carried large quantities of stone from the original rock ledges and ground them to small pebbles. In general it may be said that this glacial gravel is found in western Pennsylvania, most of Ohio, northern Indiana, northern Illinois, and in most of the northwestern States. The gravel which exists south of the glacial district, with the exception of river gravel, has been in most cases produced by a slow disintegration of the rocks in place.

THE GRAVEL ROAD

Gravel has been extensively used in certain sections of the South for road building, notably in Chatham County, Ga., of which Savannah is the county seat, and Montgomery County, Ala., while in northern Georgia, Alabama, Mississippi and Tennessee excellent roads have been built from abundant chert-gravel deposits. The gravel deposits in the South, however, are local and limited in extent, and are confined principally to the States of Virginia, North and South Carolina, Tennessee, Georgia, northern Mississippi, western Kentucky, Alabama and Arkansas. Texas is well supplied with gravel in the northeastern portion of the State. The delta regions of Mississippi and Louisiana are almost devoid of road-building materials, and the gravel deposits are small in quantity and of inferior quality. Arkansas, as a whole, is not supplied with good road material, but there are extensive deposits of gravel in the southwest portion of the State. In Kentucky the gravel is limited to local deposits along streams.

Qualities of Gravel.—Road-building gravel should possess three important qualities: hardness, toughness, and cementing or binding power. Of these three qualities the last is the most important. This binding quality is due in part to the presence of iron oxide, lime, or ferruginous clay, and in part to the angular shape and size of the pebbles composing the gravel. A good way to determine whether or not a gravel is suitable for road building is to notice its position in the pit. If the banks remain vertical after exposure to the weather, it is a reasonable inference that the material possesses a high cementing value and will cement and compact well in the road. Blue gravel is universally conceded to be the best for road construction, because it is usually derived from trap rocks. As the pebbles composing the gravel retain the characteristics which they possessed when forming part of the larger rock masses, it follows that as trap rock is considered an excellent material for road building, trap-rock gravel should occupy the same relative rank among the gravels. Limestone is, generally speaking, a soft rock, and consequently limestone gravel (which is quite rare) will usually be found soft and will wear rapidly. Quartz possesses practically no binding power, although it is a very hard mineral. Therefore, gravel which contains an exceptionally large percentage of quartz will not prove successful, unless a good binder is added. On the other hand, the chert gravels, which are composed mainly of amorphous or non-crystalline quartz, possess a very high binding value.

The shape and size of the pebbles composing the gravel have an important bearing upon its value as a road material. In order that the material may bond readily, the pebbles should be angular, and should vary in size so that the smaller fragments may fill the voids between the larger pieces. The largest pieces of gravel should not be more than two or two and one-half inches in their greatest dimensions. Otherwise, the large fragments will fail to compact and will work to the surface. On the other hand, the gravel should not be too fine, as in this case it will be equally difficult to consolidate. Many road builders consider the gradation in sizes the most important quality of road-building gravel.

The angular shape of the gravel is essential,

in order that a mechanical bond may be secured. Gravel obtained from streams is inferior to pit gravel, for the reason that the constant action of water has worn the pebbles smooth and partially round, so that it is very difficult to obtain the mechanical bond necessary in the construction of gravel roads; moreover practically all of the fine binding material has been removed by the same agency. Even if a ferruginous clay is mixed with the river gravel, the result is not likely to be as satisfactory as that obtained by the use of pit gravel.

When the gravel is taken from the pit, it should not contain more than one-fourth of its volume in sand or clay. Pit gravel frequently contains too much clay or earthy matter, while river gravel may have too much sand. In such cases, the gravel should be screened so as to eliminate the material which is too fine and that which is too coarse. The screens should have meshes of about $2\frac{1}{2}$, $1\frac{1}{2}$ and $\frac{1}{2}$ inches. Where gravel is screened in this way, it can be laid in courses on the road. The fragments which pass through a $2\frac{1}{2}$-inch screen may form the bottom course; those which pass a $1\frac{1}{2}$-inch screen, the middle course; and the fine material may be used as the top course, or binder.

Some highway engineers favour the use of clay as a binder in gravel-road construction, where the gravel requires the addition of a binder. The clay should be used very sparingly, however, as it absorbs water and causes the road to become soft and muddy. When the clay dries, it contracts and causes the road to crack. Clay is also affected by frost. Loam is frequently used as a binder for gravel roads, and consists of sand and some vegetable matter, lime, etc., mixed with clay. It possesses about the same qualities as a clay binder. The best binder of all is iron oxide, which is frequently found coating the pebbles.

Chert and Chert Gravel.—Chert is a silicious rock and occurs usually in limestone and sandstone formations. It is generally believed to be formed by a chemical precipitation from sea water. The material is found sometimes completely covering the ground; sometimes in the beds of streams and narrow valleys where it has been redeposited by the action of water; and in other cases in banks and pockets on hill and mountain sides. Bank cherts usually occur in nodular masses, but where they are found in stream beds they are often broken into angular fragments, varying in size from 1 to 6 inches. Bank cherts are easily quarried by blasting and the lumps reduced to proper size by napping hammers or by rolling.

Where these materials are found in the beds of streams they are commonly called gravel. Creek gravel formed from chert is usually of uniform size and comparatively clean, while the bank gravel often contains earthy matter and fine particles of the same material. The creek gravel wears the best, but it does not bind as readily, or form as smooth a surface as the bank deposits. Where both creek and bank cherts are available, good results can be obtained by using the former for foundation and the latter for the wearing or binding course. A road built in this way at Florence, Ala., under the direction of an expert of the Office of Public Roads, in 1898, is said to be in perfect condition at the present time, although it has never been resurfaced. Chert is found in the southern portion of the Appalachian Mountains, along the Ozark foothills, in southern Illinois, southern Missouri, northern Arkansas and eastern Oklahoma.

Gravel-road Construction.—The first step in the construction of a gravel road is to obtain the desired grade, after which the road should be given a suitable cross-section, or crown, so that the centre of the finished roadway will be from 6 to 8 inches higher than the edge of the gravelled portion for a 16-foot road. About the same ratio of height to width should be maintained for other widths than 16 feet. The subgrade should be thoroughly rolled and compacted, and all loose and unstable earth removed and replaced by sand and gravel. The gravel should then be placed on the subgrade to a total depth of from 8 to 12 inches in the centre, tapering off to a depth of from 4 to 6 inches on the sides.

Sometimes it is advisable to screen the gravel and place it in layers, and the coarser should be used for the foundation, as previously explained. The thickness of the respective courses should be approximately from 4 to 6 inches for the foundation course, from 3 to 4 inches for the second course, and from 1 to 2 inches for the surface, or binding course. Each layer should be thoroughly sprinkled and rolled with a roller weighing not less than 2 tons, and at least 2½ feet long. If a roller and sprinkler are not available, the road should be constructed in the spring, as the successive rains will cause the material to pack much better than if the road were built in the dry, hot summer or early fall. This is an exceedingly important point and one which is generally overlooked. If the gravel fails to compact, a thin layer of crushed-rock screenings applied to the surface will be found exceedingly beneficial.

McAdam condemns the practice of dumping gravel indiscriminately on the road and leaving it for traffic to compact. The following quotation taken from his report, published in 1824, applies with equal force to present-day conditions.

"The formation of roads is defective in most parts of the county; in particular the roads around London are made high in the middle, in the form of a roof, by which means a carriage goes upon a dangerous slope, unless kept on the very centre of the road.

THE GRAVEL ROAD

"These roads are repaired by throwing a large quantity of unprepared gravel in the middle, and trusting that, by its never consolidating, it will in due time move towards the sides."

The principal causes of failure in gravel-road construction may be summarised as follows:

1. Poor material.
2. Spreading the gravel in dry weather; dumping it in heaps and leaving it for traffic to compact.
3. Placing the gravel on surfaces filled with ruts and holes.
4. Insecure or poorly drained foundation.
5. Improper construction of ditches or culverts.
6. Making the road so narrow that wagons will track, thereby forming deep ruts.
7. Failure to fill ruts and holes with gravel.

The information given in Chapter IV regarding drainage applies with equal force in the construction of gravel roads, and should be followed faithfully, as otherwise a poor road will result, even if the greatest care is used in the selection of materials and in placing them upon the subgrade.

CHAPTER VII

THE BROKEN-STONE ROAD

The term "macadam" is generally understood to mean a particular type of road. That this type of construction is different from that used by John L. McAdam, and named after him, need cause but passing comment. Modern machinery and modern science have worked many changes, but the fundamental principles demonstrated by McAdam, that the foundation must be well drained in order properly to carry the loads which come upon the road; and that an aggregate of broken stone can be made to cement, or knit together, so as to be waterproof and firm enough to support traffic, still holds good. McAdam's own explanation of his method is clear, concise and to the point. In his report, published in 1824, he said:

"The roads can never be rendered thus perfectly secure until the following principles be fully understood, admitted, and acted upon: namely, that it is the native soil which really supports the weight of traffic; that while it is preserved in a dry state, it will carry any weight without sinking, and that it does in fact carry the road and the carriages also; that this native soil must previously be made quite dry, and a covering impenetrable to rain, must then be placed over it, to preserve it in that dry state; that the thickness of a road should only be regulated by the quantity of material necessary to form such impervious covering, and never by any reference to its own power of carrying weight. The erroneous opinion so long acted upon, and so tenaciously adhered to, that by placing a large quantity of stone under the roads, a remedy will be found for the sinking into wet clay, or other soft soils, or in other words, that a road may be made sufficiently strong, artificially, to carry heavy carriages, though the subsoil be in a wet state, and by such means to avert the inconveniences of the natural soil receiving water from rain, or other means, has produced most of the defects of the roads of Great Britain. . . .

"Every road is to be made of broken stone without mixture of earth, clay, chalk, or any other matter that will imbibe water, and be affected with frost; nothing is to be laid on the clean stone on pretence of binding; broken stone will combine by its own angles into a smooth, solid surface that can not be affected by vicissitudes of weather, or displaced by the action of wheels, which will pass over it without a jolt, and consequently without injury."

In addition to the modifications due to progress, other modifications of a local character must be made because of climate, topography, nature of traffic, character of the local stone, etc. Thus, while we can specify the construction for any given road to the smallest detail, it must always be borne in mind that different conditions necessarily demand changes, at least in the minor details. Furthermore, a proper recognition and appreciation of these details will invariably save money for the taxpayer.

While the macadam, or broken-stone, type of road is particularly well adapted to those carrying a moderate traffic, it is not economical as a pavement for city streets carrying heavy traffic, or on roads subjected to heavy automobile traffic, unless some special type of binder other than the stone dust is used. In some ways a macadam road resembles quite closely a well-built gravel road, but, as a rule, it will stand heavier traffic and wear better, since the mechanical bond between the aggregates is stronger than that which can be supplied by the more or less rounded pebbles of the gravel. Even in regions where gravel is abundant, a

THE BROKEN-STONE ROAD 137

macadam surfacing may prove more economical on the more heavily trafficked sections where the gravel does not furnish a sufficiently strong bond to withstand the requirements of the traffic.

Width of Surfacing.—Experience has shown that for ordinary country roads the macadam surface need not be more than from 13 to 16 feet wide, if suitable earth shoulders are built on each side. Thirteen feet allows two vehicles to pass each other safely. Sixteen feet is more satisfactory, especially when more or less fractious teams are passing automobiles. If the stone is less than 13 feet wide, there is a likelihood that the edges of the macadam will be sheared off by the wheels, unless the shoulders are made of especially good material. In fact, a width of less than 13 feet is of doubtful value, unless the surface portion is reduced to the very narrow width of 8 or 9 feet. This serves fairly well as a single track, where the prevailing loaded traffic is in one direction, and a good earth road is provided on one or both sides of the macadam.

There are many communities where during

the greater part of the year a well-kept earth road is about all that is desired, but when, for the few months during the winter or spring, these earth roads become all but impassable. Here a narrow strip of macadam with a well-kept earth road on one or both sides will sometimes answer the purpose at a much lower cost than a standard-width road of from 13 to 15 feet; for, during good weather, practically all the traffic, excepting the very heavy loads, will use the more resilient earth road, while during bad weather all will use the macadam as far as possible. The light traffic will turn out on the earth road to pass the loaded teams. Whatever may be the width of the stone, however, the shoulders should be firm enough to permit occasional passage of wheels.

In the past years it was almost the universal practice to build the macadam roads very thick. Of course, this required a very large amount of material, and made the cost extremely high. A comparatively few years ago, roads less than eight inches thick were rarely heard of, and not infrequently a thickness of at least 12 inches of macadam was thought to be necessary for a good road. To-day one of the most conspicuous means of reducing the cost has been by decreasing the thickness of the surfacing, and we find many roads supporting quite heavy traffic, although only 5 or 6 inches in thickness. Four inches of macadam after rolling is about the least thickness which is practicable, and, except in unusual cases, a depth greater than 8 inches after rolling is unnecessary.

A macadam surfacing should be hard, smooth and impervious to water. Much attention must also be given to the foundation, which should be firm and sufficiently strong to sustain any load likely to come on the road at any time of the year.

Quarrying for Material.—In opening a new quarry careful attention should be given to the inclination of bed joints or seams, which, for economical quarrying, should be parallel with and dip toward the working face of the ledge. The drainage of the quarry should also be considered and the floor level should, wherever possible, be so arranged that the water can be drained from the working face by gravity. Wherever possible the ledge should be opened where the overburden is light and where but little expensive stripping will be necessary. The quarry should be located, if possible, in such a position that gravity will assist in handling materials, so that tram-cars may carry material from the floor of the quarry to the mouth of the crusher by gravity. The loaded car in its downward trip may be made to drag the empty car back to the floor of the quarry. If the quarry is located in a pit it will be necessary to provide power for this purpose; furthermore, a considerable expenditure for pumping water will be entailed, aside from that necessary for operating tramways by mechanical power.

In removing the overburden the earth should be carried far enough away from the quarry not to interfere with future operations. When the overburden is of a tenacious nature, or when it is frozen, it may be loosened by sinking a few holes from two to five feet in depth and by charging them with explosives. Low grade dynamite is suitable for this purpose. If the overburden consists of earth or gravel, it can sometimes be removed economically by the use of water and a flume.

As road-building rocks are usually hard and tough, the drill-holes in the quarry face can be more economically placed by means of steam or compressed-air drills than by the use of hand drills. For gravels, cherts and various other soft materials used in road building, hand or churn drills may, however, be used to advantage. If a steam drill is employed, the steam may be procured from the boiler which operates the crushing plant and be conveyed to the drilling machine by small iron pipes. The quarryman should use good judgment in the selection of the positions where the holes are to be bored. He should consider the effects of the action of the explosive on the rock before him, and the relation of the bore-holes to the face of the quarry. In this connexion it is necessary that he examine carefully the fissures in the rock before the holes are drilled. To get the best results the rock should present an unsupported face on every side, but in ordinary practice this condition seldom obtains. The wall of the quarry is usually vertical, and the two free faces are the top and the breast of the rock. Ordinarily, therefore, the bore-holes should be placed as nearly parallel to the longest free face of the rock as possible.

The object in quarrying rock for road building is to shatter the materials as much as possible, and for this reason high explosives are preferred. Dynamite is a rapid and violent explosive, and produces effects very suddenly. It is, therefore, better adapted than giant powder for quarrying rock for road building. Dynamite dislodges the rock and, if properly used, reduces most of it to a size suitable for the crusher without sledging; consequently, the cost of quarrying with high explosives is cheaper than with low explosives. If giant powder is used, it is necessary either to make larger boreholes or to increase the number to obtain the same results.

It may be noted in this connexion that any explosive containing nitro-glycerin is commonly called dynamite. Dynamite is usually made by partly saturating some porous material with nitro-glycerin. The percentage of nitro-glycerin usually contained in dynamite varies from 40 to 75. If by the use of the 40 per cent. dynamite it is found that the rocks are blown out in chunks too large for the crusher, then it is advisable to use the 75 per cent. nitro-glycerin. A

THE BROKEN-STONE ROAD 143

few experimental shots with dynamite of different grades will indicate the percentage which can best be employed in any particular quarry.

Crushing.—A crusher for road building should be provided with a suitable elevator and with screens for separating the materials into their proper sizes. Revolving screens for small plants are usually about 10 feet long, 32 inches in diameter, and should revolve at the rate of about 20 revolutions per minute. The screen is divided into sections, and the lengths of each one and the sizes of holes for diabase and other harder rocks should be about as follows: First section, $3\frac{1}{2}$ feet long, holes $\frac{1}{2}$ inch in diameter; second section, 3 feet long, holes $1\frac{1}{2}$ inches in diameter; and third section, 3 feet long, holes 3 inches in diameter. No hard stone larger in diameter than will pass through the 3-inch holes in the screen should be used in a macadam road, and, therefore, stones too large to go through the larger holes should be returned to the crusher by gravity or by means of a belt conveyor, where they are recrushed. If the tailings are not recrushed, then they should be eliminated from the work. For limestones and the softer varieties of rock the size of holes in the first and second sections of the screen may be increased to ¾ inch and 2 inches, respectively. A portion of the screen which contains the ¾-inch holes should be provided with a dust jacket, as the softer rocks usually produce more dust than is necessary for binding material. The jaws of the crusher should be so set as to make as few tailings as possible, and the lengths of the screen sections should be adjusted to the same purpose.

For receiving the various sizes of crushed rocks, bins with slanting metal bottoms and sliding doors should be provided, so that the material can be loaded into wagons by gravity. Partitions should be built in the bins so as to keep the differently sized materials separated.

Two types of crushers are now commonly used in crushing rock for road building. One is the jaw type of crusher, generally used for small portable plants. In this machine one of the jaws moves backward and forward by means of a toggle joint and an eccentric, and the stone descends as the jaw recedes. As it returns, it catches the stone and crushes it. The maximum size of the products is determined by the distance the jaw plates are apart at the lower edge.

THE BROKEN-STONE ROAD 145

The gyratory crusher consists of a solid conical steel shaft supported by a heavy mass of iron somewhat like an inverted bell. By means of an eccentric, the rotary motion given to the shaft is such that every point of its surface is successively brought near the surface of the "bell," and the rock caught between the shaft and the bell is crushed. The gyratory crusher will not produce as many flat pieces or tailings as the jaw crusher, because the stones have to come in contact with two curved surfaces at the same time before they are broken. It is peculiarly adapted, therefore, to crushing rocks which are more or less laminated.

Large stationary plants are, as a rule, desirable only where the broken stone must be shipped by rail. There are several portable plants on the market which may be bought at prices ranging from about $1,500 to $2,500, and which are well adapted for country use. The complete plant includes stone crusher and engine boiler, portable bins, revolving screens and an elevator for lifting the broken stone from the discharge of the crusher into the screen. These outfits are mounted on wheels, so as to be readily moved from place to place. Where no special difficulties are encountered in setting up the plant, it may be moved from one place to another at a cost of from $50 to $100. The average output of such a plant as has been mentioned is from 75 to 100 cubic yards every day. The amount of the output, however, will depend very largely on the character of the stone which is being crushed, and the ability of the man who has the plant in charge. The hard usage to which the crusher is subjected naturally entails much repair work, and requires constant and skilful attention in order to secure the best results. Where there is a choice as to the location of the crusher, it should be placed at about the middle of the stretch of road to be built, so that the output can be hauled in both directions. The distance which the broken stone can economically be hauled will generally not exceed over one mile. This would tend to show that unless other conditions are involved, two miles of road is about all that can be economically constructed from each setting of the crusher. In general it will also be found advisable to set the crusher at the quarry and haul the crushed stone to the road, rather than to set the crusher at the road and haul the quarried stone to the crusher.

Every effort should be made to reduce the number of times which the rock must be handled. By setting the crusher at the quarry, the tram-cars can often be rigged so as to be operated, either by cable with the power supplied by the crusher engine, or by gravity, and the stone conveyed direct from the ledge and dumped on the crusher platform. With this arrangement, moving the stone will require the minimum amount of hand labour.

In some places, it may be found more economical to have the stone shipped in from some permanent crushing plant than to purchase a crushing outfit; and this feature should be carefully considered. It is well to study the character of the local stone, to ascertain whether it is such as to justify its use, or whether it would not be more economical to import a better stone, at least for the surface course.

Road-Building Machinery.—A roller operated by mechanical power has almost entirely superseded the old-fashioned horse roller. Its weight is an important consideration for two reasons: First, the cost of the roller is approximately so many dollars per ton; second, existing bridges and culverts are rarely strong enough to carry the heaviest rollers. For country roads, experience has demonstrated that a 10-ton roller is sufficiently heavy. There are a number of excellent makes of such rollers on the market, which may be had at prices ranging from about $2,000 to $3,500.

Another essential in the construction of a macadam road is the sprinkler. A sprinkler with a capacity of from 450 to 600 gallons is usually sufficient. Local conditions such as grades and the distance that water must be hauled will determine the proper size. The sprinkler should be provided with extremely broad tires, to assist in rolling the partially consolidated macadam, rather than to loosen it or to form ruts.

The road machine, or grader, is a most valuable implement, and one which is often overlooked. All too often its only use, or rather misuse, is that of scraping back upon the road the worn-out material which has been washed into the gutters. The road machine can and should be used to good advantage in shaping the road preparatory to the application of the broken stone. It is not uncommon to find that with a skilled operator the entire subgrade can be shaped with the road machine, thus doing away with considerable hand labour.

Where a large amount of road building is done automatic spreading wagons will prove economical, but since such wagons can, as a rule, be used for no other purpose, they would prove a financial burden to a contractor or a municipality that was doing but a small amount of road building.

THE BROKEN-STONE ROAD

Weight of Broken Stone.—Broken stone is frequently sold by weight. Before estimating the cost of a road, when a stone is to be paid for thus, the road officials must know how much the stone will weigh per cubic yard. The erroneous impression that all stone weighs the same per unit volume is quite general throughout the United States. One often hears it stated that a cubic yard of broken stone weighs a ton and one-third, regardless of the kind of stone. The following table taken from Farmers' Bulletin No. 338, United States Department of Agriculture, gives the specific gravity and weight of a number of the more common rocks:

SPECIFIC GRAVITY AND WEIGHT OF VARIOUS ROCKS.

Number of samples tested.	Name.	Specific gravity.			Weight per cubic foot of solid rock.			Weight[1] per cubic yard of solid rock.		
		Max.	Min.	Av.	Max. Lbs.	Min. Lbs.	Av. Lbs.	Max. Lbs.	Min. Lbs.	Av. Lbs.
8	Peridotite (trap).	3.55	3.25	3.40	221	203	212	2,984	2,741	2,862
124	Diabase (trap)...	3.20	2.60	2.95	200	162	184	2,700	2,187	2,484
33	Diorite (trap)...	3.35	2.70	2.85	209	168	178	2,821	2,268	2,403
60	Schist	3.20	2.65	2.90	200	165	181	2,700	2,227	2,448
11	Felsite	2.80	2.50	2.65	175	156	165	2,362	2,106	2,227
53	Quartzite	3.10	2.50	2.70	193	156	168	2,605	2,106	2,268
358	Limestone	3.10	2.00	2.65	193	125	165	2,605	1,687	2,227
106	Granite	3.00	2.00	2.65	187	125	165	2,524	1,687	2,227

[1] Tons of 2,000 pounds.

The above table gives the weights of the solid rock as it is found in the quarry. If it is assumed that the volume of the stone, after it is crushed and lies in the bins, has a void of 50 per cent., and the average weight of peridotite is compared with the average weight of granite, it will be seen that the crushed peridotite weighs 1.43 tons to the cubic yard, while the granite weighs only 1.11 tons. The heaviest diorite weighs 1.41 tons to the cubic yard, and the lightest only 1.13 tons. Differences as marked as these emphasize the great need of careful determination of the weight of the material before any contracts are let.

When broken stone is purchased by measurement, from cars or in wagons, the specifications should always state where the measurements are to take place. It is evident that the stone will occupy considerably more space when it is first loaded into either the car or the wagon than after it has been jolted about in transportation, either by rail or on the wagon road.

Earth Work.—No earth work should be undertaken until the grades have been definitely established and the grade stakes set. Such work of course belongs to the engineer, and this holds true whether it be in regard to the construction of a new road or the reconstruction of an old one. In fixing the grades, care should be taken to adjust the cuts and fills so that there will be no undue amount of waste or borrow as explained in Chapter IV. No extreme refinement, such as is sometimes practised on railroad work in balancing the cuts and fills, is necessary in highway construction. In most States, the right of way provided is wider than the width necessary for the roadway. Therefore, where more fill is needed, the additional material can readily be secured by simply widening the adjacent cut to the desired extent, and, where the cuts are in excess, convenient wastage can readily be found by simply widening the adjacent fills. This does not mean, however, that the work is to be done in a haphazard manner, but that the computations and surveys shall be carefully made and where additional widenings are needed they shall be immediately staked out on the ground.

It is obvious that the subgrade, or foundation, of a road is the part most nearly permanent.

The grades should, therefore, be studied most carefully, since they cannot be changed without great expense.

Drainage.—Drainage is absolutely essential to macadam, as well as to any other form of road. The road should be so constructed that it will shed the water to the side ditches as rapidly as possible, and the side ditches in turn must be of such size and slope as to remove the water from the road quickly and completely. For a narrow macadam road, a crown or side slope of ¾ of an inch to the foot for the macadam portion will be about right. For a wide road this will give too much crown, and the side slope must be reduced to ⅝ or perhaps ½ inch per foot. The slope of the shoulders should be equal to, or perhaps in general, a little greater than that of the macadam. The slope of the side ditches must be made to vary somewhat with local conditions. If possible the slope should be sufficient, so that the ditches will be self-cleansing, and not have a tendency to fill with detritus washed from the road. On the other hand, the slope should not be so great as to cause erosion. Where steep grades cannot be avoided, the gutters or side ditches must be either paved, or else stops placed at occasional intervals to check the velocity of the water. Nor should the practice, which is so often found, of carrying the water along the road for long distances, be tolerated. Water is always an element of danger to a road, and should be gotten rid of as quickly as possible. Every outlet should be utilised for this purpose, even though it involves the construction of a few more cross-drains.

Surface water is not the only danger to a macadam road. In many places special attention must also be given to the underground waters. It is sometimes possible to drain the road with open side ditches, but deep ditches on the roadside are an element of danger, and, where the ground waters are to be removed from the road, it will usually be preferable to employ tile drains. Sometimes the direction of the movement of the underground water is such that a single drain on one side of the road will be sufficient. In other cases, a drain will be required on both sides. The best practice in road drainage is to remove the ground water to such a depth that there will be no danger from the heaving action of the winter frosts. The drains usually consist of narrow trenches filled more or less completely with broken stone or gravel, and having a drain tile near the bottom. The tile used is ordinarily the open-joint drain tile, which must be laid true to grade, and provided with free outlet. Sometimes the pipe is omitted and the trench is filled entirely with stone, when it is called a blind drain. This practice, however, is not to be recommended where large quantities of water need to be removed at any time during the year.

Cross-drains may be made of concrete, or, if not large, iron pipe or vitrified clay tiles may be used. Recent improvements in the manufacture of non-corrosive steel have made that material available for this purpose. Vitrified clay tile has also been used, as well as abused, to a large extent in past years. It should never be laid close to the road surface, nor where there is the least danger of the drain ever clogging in cold, winter weather. Many failures of clay tile have been caused in the northwestern States by a winter thaw during which the tiles become clogged with slush, ice and water, and then this thaw is followed by a freeze, which, of course, bursts the tile. Of all materials at present, concrete seems the most durable, as well as, in the majority of cases, the most economical. Where large bridges or culverts are required, detailed designs should always be made before construction.

Subgrade of Macadam.—The surface upon which the broken stone is to be placed must be hard, smooth and carefully crowned. This is necessary to prevent excessive use of stone on the one hand, or the undue waste of stone on the other. If the foundation is not hard and firm, the stone will be pressed into it by the roller, and thus wasted. If it is not properly crowned, an unnecessary quantity of stone will be required. When macadam is to be of uniform thickness throughout its cross-section, the crown of the subgrade must be the same as that of the finished roadway. If the macadam is to be thicker at the centre than at the sides, a part of the crown will be of the macadam itself, and the centre of the subgrade should be raised only enough to produce the surface crown when the stone is in place. As has already been stated, the road machine is a most useful implement in shaping the subgrade.

After the roadbed is shaped to the approximate cross-section, it should be rolled until it is hard, firm and smooth. If soft places are found, or if depressions develop during the rolling, these should be filled with good material, and then further consolidated with the roller until the subgrade has the required cross-section as nearly as practicable.

Placing the Stone.—The stone should be placed in courses not to exceed 6 inches in depth when loose, as this is about the greatest depth which can be thoroughly consolidated with a roller. On the prepared subgrade, which has been properly rolled and consolidated, is spread the first course of stone, usually varying in size from $1\frac{1}{2}$ inches to 3 inches in the largest dimensions. Much larger stone than this should not be used in the foundation unless the road is to be very thick. In practice two methods are used for spreading broken stone. One is to dump the stone on a board platform and then shovel it into place on the road. The other

THE BROKEN-STONE ROAD 157

is to use either an automatic spreader, or else dump the load directly on the roadway and simply spread it by pushing a portion of the stone in the different directions, or until the required thickness of loose stone is obtained. When the stone is spread by simply raking off the top of the loads dumped directly on the roadway, the proper consolidation is not secured by rolling; the stone will be denser and more compact where the load is dropped. An uneven roadway sometimes results, and in some extreme cases the position of each load can be clearly seen after the road has been in use for some time. To obtain the best results each load of stone should be dumped in three or four piles. This facilitates the spreading and insures a more uniform distribution of the material.

When about 100 feet or so of the first course have been spread, the rolling should begin. The roller should commence on the outer edge of the macadam with the outer wheel well up on the shoulder, and gradually work towards the centre of the roadway. When the centre has been reached, the road should be crossed over, and the other side rolled in the same manner as the first. After both sides of the roadway are moderately firm, the roller should be moved gradually towards the centre, until the entire lower course is thoroughly compacted. Where the foundation is poor, or a bad silt soil is encountered, it is well to use a filler in the bottom course. This should consist preferably of a good dry sand which is spread over the stone after it has been rolled fairly well. The rolling is then continued until the voids have been forced completely full of sand or stone screenings. No clay, loam or perishable foreign material should be allowed in the filler. Not only will a filler prevent a slippery clay from working up into the interstices of the stone, but it will also assist in consolidating a stone which does not possess good mechanical bonding qualities, such as quartzite.

If depressions develop as a result of the rolling, additional stone of the same size used in the course should be added and the rolling continued, so that before the second course is applied the lower course is smooth and true to cross-section.

THE BROKEN-STONE ROAD 159

After about 100 feet of the first course of stone is rolled, the second course, consisting of stones varying in size from 1½ inches down to ½ inch, is spread and rolled in the same manner as the lower course. The thickness of the second course usually varies from two to four inches compacted. The stone should be carefully spread and considerable vigilance is necessary if the spreaders are not accustomed to their work, in order to prevent the surface having a wavy appearance when the rolling is completed. It is quite a temptation with the workmen to fill these small depressions with screenings rather than with the stones of the proper size.

When the surface is thoroughly compacted, which is usually judged by the absence of any wavy motion in front of the roller, the screenings or binder course is applied. Only sufficient screenings should be applied to fill the voids in the stone and form a very slight covering on the surface. Screenings should be spread in successive thin coats with alternate rolling. Sometimes it is a good plan to pass the roller once or twice over the screenings as they have been spread on the roadway while they are dry. The sprinkler is then put on in advance of the roller, and as much as possible of this dust of the screenings is flushed into the crevices of the stones. The sprinkling and rolling should continue until the surface puddles, showing that the voids are substantially filled. The process of binding the top course is the most critical one of the entire job. The ability of the roller operator is a very important factor in macadam work. The appearance of the road surface depends to a large extent on his skill.

As soon as the road has been puddled, it should be allowed to dry a few days, and may then be opened to traffic. In fact, if a road can be opened in sections as completed, it is more preferable than to wait until the entire road is done, and then throw it open. Where it is opened in sections, it will be found possible at times to run back over it with the sprinkler and roller. Traffic on a green road always produces more or less roughness or even ravelling, so that permitting travel on the road while it can still occasionally be reached with the roller, is one of the most rapid ways of obtaining the final set to the road.

On a very clayey or silty soil considerable care must be exercised in order to prevent the water from reaching the subgrade in quantities sufficient to soften it. If much water reaches the subgrade, there is great danger that the clay will be forced up into the stone, and depressions will result, and the undesirable clay will penetrate into the stones.

Cost of Macadam.—No formula has yet been devised whereby the cost of macadam roads can be computed for any locality without a detailed survey and close examination of conditions. There are too many uncertain factors which enter into the construction of the road in such varying proportions to make a table of cost of macadam roads of any great practical value. Each road is a problem in itself, and while one mile of road may cost a given amount, the adjoining mile may often cost twice as much, for no other reason than the variation of necessary factors. Roughly, it may be stated that in various parts of the United States the cost of macadam roads, having a width of 15 feet and a thickness of seven or eight inches, ranges from $2,000 to $10,000 per mile.

CHAPTER VIII

SELECTION OF MATERIALS FOR MACADAM ROADS

It is impossible to construct a satisfactory macadam road with inferior materials. If a very soft rock is used, the road will wear rapidly and soon have to be renewed. If the rock does not possess sufficient binding power and no adequate binder is used, it will not consolidate, and the road will soon go to pieces. Enormous sums of money have been wasted through the use of unsuitable materials, and there are many examples of unnecessary expense through the use of material brought from a long distance when one locally available would have answered the purpose equally well.

It has been found that, in a general way, certain classes and types of rock are more suitable than others for road building. For example, trap rock is considered to be an excellent material for macadam roads, while quartzite is of very little value except in the foundation. Unfortunately, the trap stones are not common to all sections of the United States. Some of the fine-grained granites usually give good results, as do the felsites, some of the harder limestones and the dolomites.

In general the micaceous, schistose and metamorphic rocks have but little value as surfacing material. Sometimes, however, the harder of these may be used for the lower course of the macadam, while the upper course is built of a better grade of stone. Some of the coarsely crystalline granites and some of the limestones, if very soft or if crystalline to any extent, are of very little value. On the other hand, however, there are instances recorded where certain schistose rocks have been used with excellent results. The following table gives in compact form the classification of all rocks used in the construction of macadam roads:

SELECTION OF MATERIALS

GENERAL CLASSIFICATION OF ROCKS.

Class.	Type.	Family.
I. Igneous	1. Intrusive (plutonic)	a. Granite b. Syenite c. Diorite d. Gabbro e. Peridotite
	2. Extrusive (volcanic)	a. Rhyolite b. Trachyte c. Andesite d. Basalt and diabase
II. Sedimentary	1. Calcareous	a. Limestone b. Dolomite
	2. Siliceous	a. Shale b. Sandstone c. Chert (flint)
III. Metamorphic	1. Foliated	a. Gneiss b. Schist c. Amphibolite
	2. Nonfoliated	a. Slate b. Quartzite c. Eclogite d. Marble

Igneous or fire-formed rocks are those which at one time have been in a molten state and have solidified, either underground or on the earth's surface. Heat, pressure, and the chemical composition of the rock, together with the presence of vapours, were the causes which governed the final structure of the material. Those rocks which were consolidated deep underground are known as plutonic and are formed of coarse crystals. Examples of plutonic rocks are granite, syenite, and diorite.

The rocks which have solidified at the surface include rhyolite, andesite, and basalt. The colour of igneous rocks varies from light grey, pink and brown to dark steel grey or black. The dark varieties are generally called trap, a term derived from *trappa,* a Swedish word meaning stair, as the formation frequently resembles stairs.

Sedimentary rocks are composed of fine rock particles and fragments which have been produced by the disintegration of rocks of various types, carried by running water and deposited in layers on sea or lake bottoms. Examples of sedimentary rocks are limestone, sandstone and shale.

Metamorphic rocks are those which have been formed by the action of chemical or physical forces on igneous and sedimentary rocks. Examples of this class are gneiss, slate, quartzite and marble.

As far as it is possible to determine the relative value of the various rocks for road building according to their mineral classification, it may be said that the following is the order in which they should be ranked:

SELECTION OF MATERIALS

1. Trap.
2. Syenite.
3. Non-crystalline Limestone.
4. Chert.
5. Granite.
6. Mica Schist.
7. Quartzite.

Stone from a ledge, because of its uniformity, is usually better than field stones, but if the ledge is of an inferior grade of rock it should not be used merely because it is ledge stone, in preference to field stones of better quality. The aim in the selection of a road material should always be to get a rock of uniform quality. Badly weathered stone from the surface or outcrop of a ledge should never be mixed indiscriminately with the fresh stone from the interior.

Physical Qualities.—The mineral classification of rocks is by no means a conclusive test of their fitness for road building, as there is a wide variation in the qualities of different outcrops and deposits of materials belonging to the same class and type.

It is the aim of the road builder to obtain a road with a surface as nearly smooth as possible, not too hard, too slippery, or too noisy, and which will be as free as possible from mud and dust. These results are to be obtained and maintained at as small a cost as possible. In order to produce even approximately such conditions, it is necessary that only rock possessing certain essential physical qualities, irrespective of mineral properties, be used.

In order to determine what qualities are essential in a road material, it must be borne in mind that the road will be called upon to withstand the wearing action of wheels and horses' hoofs, as well as the action of the elements, in the form of rain, wind and frost.

Hardness is the quality possessed by rock which enables it to resist the wearing action of the wheels and horses' hoofs. It is evident that hardness is an essential quality, and particularly so if the road is heavily travelled.

Toughness is that quality in the rock by which adhesion between the crystalline and fine particles of the rock is so great as to give it power to resist fracture when submitted to the blows of traffic. Its quality is different from hardness. The difference is illustrated by the statement that the resistance by rock to the grinding of an emery wheel would be considered hardness, while the resistance to fracture when the rock is struck by a hammer is toughness.

A third and very important quality in road material is the *cementing* or *binding power*, which is the property possessed by rock dust to form a cement or bond when wet, whereby the coarser fragments of the surface course are bound together and the whole forms a smooth, water-proof shell or crust. Since it is absolutely necessary to protect the subgrade from water, it will readily be seen that the rock which does not possess sufficient binding power is likely to form a loose surface, which will permit the water to sink through and soften the subgrade or foundation, thereby destroying the stability of the whole road.

It is important in the selection of material for a macadam road to consider the character of traffic which the road will be called upon to sustain. To make this point clear, the theory upon which the macadam road rests may be again explained as follows: The rock dust which

170 ROADS, PATHS AND BRIDGES

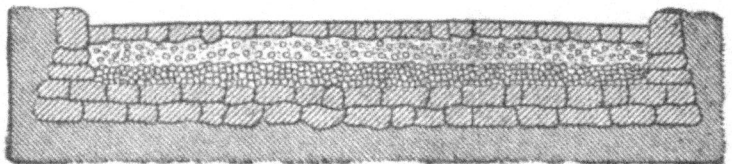

Cross Section, Roman Road (Appian Way).

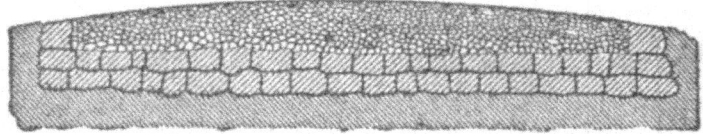

Cross Section, French Road (Roman Method), previous to 1775.

Cross Section, Trésaguet Road, 1775.

Cross Section, Telford Road, 1820.

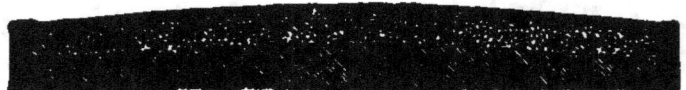

Cross Section, Macadam Road, 1816.

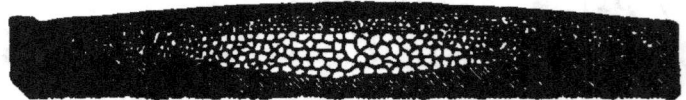

Cross Section of Modern Macadam (Massachusetts) Road with V-shaped foundation.

Cross Section of Modern Macadam Road.

SELECTION OF MATERIALS 171

fills the voids between the angular fragments of stone and forms a cement or binding material when wet is gradually carried away by wind, rain and the action of traffic. It is known, however, that the hoofs of the horses and the iron-tired wheels of vehicles wear a sufficient amount of new dust from the fragments of rock to replace that which is lost in this manner. Consequently, the bond of the road is automatically renewed. If, therefore, a very hard rock is used in the construction of a macadam road, heavy traffic will be necessary in order that there may be sufficient wear to produce the essential rock dust. If, on the other hand, a soft material is used for a heavy traffic road, the rock will be worn away far more rapidly than is necessary for the automatic binding of the road. Practical road builders realise that for very heavily trafficked roads a hard material, such as trap rock, is essential, and that for light trafficked roads limestone or other material which is not as hard as trap rock will serve every purpose.

An important series of experiments conducted by the United States Office of Public

Roads developed the fact that the addition of limestone screenings to hard material, such as granite or diabase, increases the cementing quality to a marked degree. The experiments were carried still further and lime water was mixed with the granite. The tests showed in every case a marked increase in the cementing value of the granites treated, and the inference is that the addition of lime will greatly increase the binding power of certain road-building rocks. As a result of these experiments, the conclusion was reached that mixtures of acid and basic rocks give a higher cementing value than either rock alone.

To enable the road builder to determine the value of a rock as a road material, a number of tests have been devised. These tests, however, require special apparatus, and much skill and good judgment on the part of the operator. The Office of Public Roads of the United States Department of Agriculture maintains a splendidly equipped laboratory where tests and analyses of rocks are made free of charge. No construction of any importance should ever be undertaken without having the rock tested, un-

SELECTION OF MATERIALS 173

less it has already been subjected to that best of all tests, actual use on the road for a number of years.

Directions are issued by the United States Office of Public Roads for the selection and shipment of specimens of road material for laboratory tests, and if followed carefully, the selection of the best available material should be insured. In order to have road materials tested in the laboratory of the Office of Public Roads, the instructions below must be carefully followed:

1. All samples should be selected to represent as nearly as possible an average of the material.

2. A sample of rock for laboratory tests must consist of stones which will pass through a three-inch but not through an inch and a half ring—*excepting one piece,* which should measure, approximately, four by six inches on one face and be about three inches thick. The whole sample should weigh *not less than thirty pounds.* It is desired that samples of rock be shipped in burlap bags.

3. A sample of gravel must weigh not less than twenty-five pounds, and should not contain stones over one inch in diameter. Such samples must be shipped in boxes, sufficiently tight to prevent the finer material from sifting out.

4. A blank form and addressed tag-envelope will be supplied by the Office for each sample. The blank form must be filled and placed in the tag-envelope, which must be used as the address for the sample. It is essential that the blank form be filled with the utmost care, as they are filed as records of the samples.

5. The Office desires to keep a record of the actual wear on roads built of the materials tested. If the material which this sample represents has been or is about to be used on roads, this Office would desire to be informed of the addresses of those in charge of the construction and maintenance of such roads.

6. Samples must be shipped, freight or expressage PREPAID, and bills of lading or express receipts forwarded by mail to the Office of Public Roads, Department of Agriculture, Washington, D. C.

7. The Office makes no charge for tests.

Distribution of Road Materials. Trap Rock.—Trap rock is abundant throughout the most of New England, except in the northern part of Maine. The best quality is found in the valley of the Connecticut, south of the Vermont and New Hampshire line, and along the coast between Boston and Eastport, Me. Excellent trap is found in the upland portion of New Jersey and in parts of Maryland and Pennsylvania. New York is not so well supplied except along the Hudson in the vicinity of the Palisades. South of the Potomac River trap rock is limited to the Blue Ridge Mountains and to the Piedmont country east of the Appalachian Mountains. In the country between the Appalachian Mountains and the Mississippi River, very few trap dykes occur. The northern part of Michigan is abundantly supplied with trap rock. West of the Mississippi, in southern Missouri, Arkansas and Oklahoma, there are a few scattered rocks of this nature, but in the Rocky Mountains, and on the Pacific Coast, excellent trap rock abounds.

Granitic Rocks.—Granitic rocks, which include granites, syenites, and the harder gneiss, follow in general the same distribution as the trap rocks, and it is said that between the traps and granites about one-third of the area of the United States is well supplied with road-building stone.

Quartzites.—The quartzites are found particularly in the mountainous districts of the Appalachian and Cordilleran regions, and in the Ozarks and Adirondacks.

Limestones.—Limestone is found in many parts of the Mississippi Valley, in the southern parts of Indiana, Ohio, the Valley of Virginia, in Kentucky, eastern Tennessee, and northern Alabama.

CHAPTER IX

MAINTENANCE AND REPAIR

THE terms maintenance and repair are very frequently used as synonyms, but there is a wide distinction between the two operations. *To maintain a road means to keep it always in good condition, while to repair a road means to make it good only occasionally.* In other words, repair sets in after maintenance fails to keep the road in proper condition. To maintain a road, therefore, means not to let it become bad; to repair it, means to improve it after it has become bad.

There is no phase of the subject of road improvement so important, and which is so often neglected, as that of maintenance. Roads may be constructed in a most scientific manner, and out of the best materials available, but unless they are properly maintained, they will sooner or later go to pieces. On the other hand, roads may be very poor, but with systematic maintenance and repair, they may be rendered passable at all seasons of the year for ordinary traffic. No road has ever been so well constructed that it did not need to be maintained. Even the tremendously massive roads of the Romans have almost disappeared owing to this lack.

It has been the universal practice in America to repair the roads at such times as will interfere least with individual duties, and this has crystallised into repairing the roads once or twice a year. So hard and fast has this custom become in many of the States that, even if costly macadam roads are constructed at great expense, they are allowed to go to ruin because minor defects are permitted to go unrepaired until they result in practical destruction of the road.

A road is no more than completed before the destructive forces set in. These destructive agencies are largely due to traffic and the elements. They act and react upon each other in such manner as to make the determination of the wear due to each a very difficult matter. It has been estimated that ordinarily about 80 per cent. is due to traffic and 20 per cent. to weathering. Of the former about 56 per cent. is believed to be due to the effects of the horses' feet, especially the calks, and 44 per cent. due to the abrasion of the wheels. Ordinarily the forces of destruction may be given in the order of their importance, as the shoes of the horses, the wheels of the vehicles, and the weather.

Even the most superficial examination of our roads tells us that the wear on our highways is no negligible amount. The hardest rock will wear, and the most important road problem before highway engineers to-day is one of road maintenance, rather than of road construction. It is worse than folly to build expensive roads and then expect them to take care of themselves. Not a few States are awakening to the sad realisation, hastened, to be sure, by the automobile, that even State-aid roads must be maintained after they are built.

No more admirable system of maintenance could be devised than that which is followed in France. Every mile of road is inspected daily, and the slightest defect is mended at its inception. The maintenance-of-way departments of our great railroad systems do not provide a more thorough inspection of railroad tracks than do the French for their public roads. The changes which should come in the American system will mean the adoption of a continuous system of repair and a methodical inspection of all roads.

American Methods of Maintenance.—There are three systems of road maintenance in use in this country, viz.: the contract system, the labour-tax or personal-service system, and the system which provides men permanently employed to look after particular sections of road.

The contract system has been used to some extent in various States, but it has never been found entirely satisfactory. As a general rule, the amount paid for this work is small and such poor service is rendered that in many cases the roads have become worse rather than better. Some of the European countries adopted it during the last century, but the experiment proved a failure.

The working out of personal or property taxes upon the public roads has never proved satisfactory. No State or community has ever built or kept in repair a system of first-class improved roads under the personal-service or labour-tax system. In fact, this system is not applicable even to earth roads. Its principles are unsound, its operations unjust, its practice wasteful, and the results obtained under it are unsatisfactory in every particular.

Undoubtedly the best system of maintenance is that which provides for the permanent employment of skilled labourers or caretakers, who may have charge of particular sections of road or who may be assigned to any part of a county or district where the work is most needed. Men employed in this way become experts in their particular line of work, and if they make mistakes one year, they are pretty apt to correct them the next; but, under the labour-tax system, these mistakes are repeated continuously. If one man is employed to look after a particular stretch of road, or to do a particular class of work, he will soon learn to take pride and interest in his work.

This system has been adopted in this country only to a limited extent. It has been used by the Massachusetts Highway Commission for several years. The New York State Highway Commission introduced it in the year 1910 for the maintenance of State roads, and Allegheny County, Pennsylvania, employs it for the maintenance of about 100 miles of county roads.

While it would be manifestly impossible to adopt this system throughout the entire country on account of limited resources and sparse population, still it is believed that there are many places where it might be used with great success. It would be difficult to find a county which is so poor that it could not afford to employ continuously eight or ten labourers and three or four teams to maintain and repair its roads; and many counties could well afford to employ ten times such a force. That such a plan would be more effective than either the labour-tax or the contract system would appear to be self-evident.

Neglect of Earth Roads.—Of all our roads, the earth roads are probably the most neglected. Experience has shown that by proper maintenance a well-constructed earth road can be transformed into something better than elongated mud-holes. The first and last commandment in the maintenance of earth roads is to keep the surface well drained. Water is the great enemy to our clay and heavy-soil earth roads, and must be removed immediately, or much mud is the result. To insure good drainage, the ditches must be looked to and obstructions removed, and the smooth, raised crown of the road maintained. For this purpose the split-log drag, or some similar device, is very useful, and at the same time inexpensive.

The drag should be used while the road is wet from recent rain and while the clay is plastic and too wet for the use of a road machine, but not in such a state as to adhere very much to the drag. The theory of the drag is simply this: Most clays and heavy soils will puddle and set very hard. The drag is essentially a puddling machine, and hence must be used while the earth contains enough moisture to puddle. The drag should be driven up one side of the road and down the other, inclined at an angle of about 45 degrees to the line of the road, so that a little earth is always moved toward the centre. In this way, the crown will be maintained, ruts and depressions filled, and the entire surface plastered over with a thin coat of puddled clay or earth, which packs very hard under passing traffic. The drying action of the sun and wind bakes the surface into a hard crust. Continued use of the drag will soon cause the road to be literally shingled over with successive layers of puddled earth as hard and dense as earth can be made without costly treatment.

The following points should be borne in mind in dragging a road:

1. Make a light drag which is hauled over the road at an angle so that only a small amount of earth is pushed to the centre of the road.

2. Ride on the drag and never drive faster than a walk.

3. Begin on one side of the road or wheel track, returning on the opposite side.

4. Drag the road as soon as possible after every long wet spell, when the mud is in such a condition as to puddle well and still not adhere too much to the drag. A few draggings on any given road will give the operator a clue to the proper way and best time to drag.

5. Drag at all seasons of the year, but do not drag a dry road. If a road is dragged immediately before a cold spell, the road will freeze in a smooth condi-

MAINTENANCE AND REPAIR 185

tion and do away with our extremely rough winter roads.

6. Always drag a little earth toward the centre, with the aim of keeping the slope of the crown from ¾ inch to 1 inch to the foot. If the drag cuts too much, shorten the hitch or change your position on the drag.

The best results from dragging are obtained only by repeated applications. One or two annually will not maintain an earth road in its best condition, unless the traffic is light. Some gravel roads may be considerably improved by dragging, especially if the gravel contains any clay, but it will do no good on a well-bonded macadam road.

The Sand-Clay Road.—The best method of maintaining a sand-clay road is by means of a split-log drag or a reversible road grader. The small ruts and depressions which are liable to form under heavy traffic, particularly in wet weather, should be filled as soon as possible after they are formed; otherwise the traffic is liable to cut through the sand-clay surface and destroy whole sections of the road, making it necessary to resurface it. The drag or road machine should be used in damp weather so that the surface will pack and bake while the road is drying out.

If the surface becomes loose in dry weather, this is an indication that there is not enough clay in the mixture. This defect may usually be remedied by a thin application of clay, raked in with a tooth harrow or worked in by means of a disc harrow. The mixing should be done in damp weather, and just before the road dries out it should be scraped with a reversible grader or a split-log drag.

If the road becomes sticky or muddy in wet weather, this indicates that there is not enough sand in the mixture. A thin layer of sand applied in wet weather will usually remedy this defect. The sand can ordinarily be worked into the surface by traffic, although quicker results can be obtained by "discing" or harrowing. Small holes and depressions in the surface may be remedied usually by the application of a small quantity of sand and clay of the proper mixture raked into position with a garden rake. These patches should be carefully made so that when they finally consolidate, the surface will be smooth and free from bumps or depressions.

MAINTENANCE AND REPAIR 187

Care of Gravel Roads.—A gravel road requires more attention the first year after its construction than for many years thereafter. Small ruts and depressions should be filled as soon after they are formed as possible; otherwise, they will catch water, which soaks through to the foundation, softens the subgrade, and causes the whole surface to wear rapidly or to give way entirely. A small quantity of material will fill incipient ruts and holes, but if neglected, a cartload of material may be required to repair a hole, which might otherwise have been filled by a shovelful.

Small depressions may be filled by adding fresh gravel, but, as a general rule, all that is needed is to rake the loose gravel from the side of the road into them. A split-log drag or some similar device is very useful for this purpose. If fresh gravel is added, all coarse material should be eliminated and the gravel should contain enough fine material to cement it together. A little clay is sometimes helpful, but too much clay will render the road dusty in summer and muddy in winter.

A gravel road should not be considered finished until it has been in use for at least one year. If the road has been properly maintained, it will be found, after about a year's service, that the wheels of heavily loaded wagons will not form ruts or depressions in the surface. The road will require but little attention for several years after it has passed through this formative period. Attention should still be given, however, to the side ditches and to culverts. They should be kept open and free, so as to permit water to drain quickly away from the road, especially during the spring of the year, when the snow and ice are melting. After a few years it will be found that the gravel will work toward the sides of the road, leaving a depression in the centre, which will prevent or interfere with the flow of the water from the surface to the side ditches. This will not be the case, however, if the road is dragged with a split-log drag, or surfaced from time to time with a reversible grader.

If the reversible grader is used for this purpose, care should be taken not to shove earth, sods, or weeds from the side ditches to the centre of the road, or if this is done, then such trash

MAINTENANCE AND REPAIR 189

should be, by all means, removed. Hundreds of miles of fairly good gravel roads are seriously injured every year by this practice of piling sods and trash in the centre.

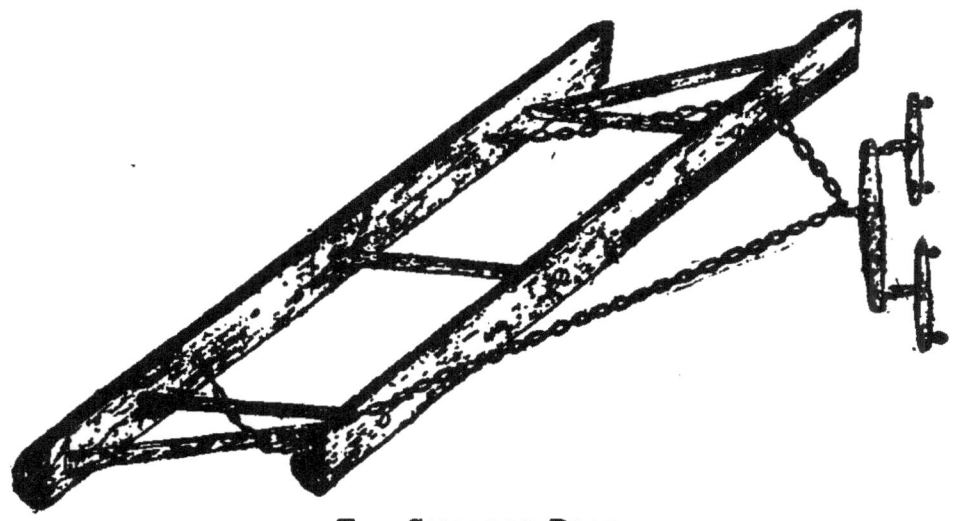

THE SPLIT-LOG DRAG

The best time to use the drag or road machine on a gravel road is just after a heavy rain, when the surface is comparatively soft and when the material which is scraped towards the centre will pack again into a hard crust. This work should never be done in dry weather, for the reason that the loose material will soon turn to dust or mud.

The crown of the average gravel road should be maintained at about an inch or three-quarters of an inch to the foot. That is, a road which is 18 feet wide from shoulder to shoulder should have a crown of not less than 6 inches nor more than 9 inches. If the crown be greater than 9 inches, the traffic will be forced to use the centre of the road, which will soon cause ruts or depressions to form in the surface. If the crown is less than 6 inches, the surface will not properly drain itself. By making the slope about three-quarters of an inch to the foot, it will be found that the traffic will use the whole surface of the road and will, in that way, distribute the wear much better than with a higher crown.

If gravel roads are neglected, especially while they are new, they will soon go to pieces and the money and labour expended upon them will be wasted. Constant attention should be the watchword. If it is estimated that a 5-mile stretch of road will require a hundred days' labour each year to keep it in repair, then it is much better to distribute that labour throughout the year than to have the work done all at one time. One man can do better work in maintaining a gravel road by working 313 days an-

MAINTENANCE AND REPAIR 191

nually than 313 men can in working one day annually. The old adage, "A stitch in time saves nine," applies with equal force to the maintenance of a gravel road.

Even though a gravel road may be maintained in good condition, it will require resurfacing from time to time, especially if the road is heavily travelled, or if the material is poor. For this repair work the very best gravel available should be used, and the work should, if possible, be done when the ground is damp, so that the new material will knit and bond itself to the old road surface.

It is the usual custom in many communities in repairing an old gravel road to dump wagon-load after wagon-load of material in a windrow in the middle of the road, and then to leave it in that condition to be spread by the traffic. This practice cannot be too severely condemned. These large piles of gravel in the middle of the road are dangerous, especially at night to those travelling in buggies or automobiles, and many serious accidents are due to this cause. The material if piled up in this way is gradually pushed by the traffic towards the side ditches, and by the time the road is consolidated, at least 50 per cent. of the gravel is wasted by being ground and pounded by the wheels of vehicles, the hoofs of horses and the tires of automobiles. Even after the road is consolidated, the surface is full of bumps and holes, which render it disagreeable to travel and difficult to maintain.

The best practice in repairing is to dump each load of gravel in three or four places and then to pull the material into position with a garden or other suitable rake, eliminating all pebbles larger than two inches in diameter. Another good practice is to spread the material over the surface with a reversible grader and then rake it with a tooth harrow. In no case, should the load from one wagon be dumped in one place, as this produces a bumpy surface.

The repairing should by all means be undertaken in the spring of the year, so that the gravel will have time to consolidate before dry weather sets in. If the gravel is spread in the summer or early fall, it will remain loose until the winter rains come. The water will then penetrate to the foundation, rendering it so soft

MAINTENANCE AND REPAIR 193

that much of the gravel will disappear during the ensuing winter.

Where extensive repairs are to be made, a steam roller with spiked wheels may be used to good advantage in tearing up the old roadbed. A tooth harrow may then be drawn over the surface, which will permit the dirt, clay and sand to sift to the bottom and will bring the loose gravel to the surface. The road may then be rolled, and a layer of suitable binding gravel applied, after which it should be sprinkled and again rolled until it is ready for traffic. The roller and sprinkler will not be needed if the materials pack well under traffic, although better results can usually be obtained by their use.

Some gravel roads may be considerably improved by surfacing them with a thin layer of hard, tough rock screenings, such as the traps and better grades of granite. This method has been pursued for parks and boulevard roads in the District of Columbia. A large mileage of gravel roads has been surfaced with trap-rock screenings. These roads have the appearance of macadam and wear practically as well, and at the same time are much cheaper than if they were built entirely of broken stone. Some of them have been oiled recently and with very gratifying results.

Maintaining Macadam Roads.—The causes of wear on macadam roads are the weather, the wheels of vehicles and the hoofs of horses. The weather acts to some extent directly on the materials, but to a much greater degree indirectly. Frost is one of the most active of these agencies. The expansion and contraction caused by frost leads to a general disintegration of the road surface. This is especially true where much clay was allowed in the binder, where the road surface was porous or the drainage poor. When such a road thaws out after a hard freeze, the macadam will practically be a layer of loose stones into which the traffic will cut, forming ruts. Rain, following a frost and thaw, is especially damaging, and a series of thaws, rains and frosts, will entirely destroy the bond in a road when once the water has gained access into the stone. Frost has but little, if any, effect on a dry, well-kept road. The solution is self-evident. Look after the drainage very carefully in the fall and be sure that the surface is as nearly waterproof as possible, so that the road will, at the beginning of winter, be dry and not full of water. Violent rains on exposed localities wash out the binder, and sometimes the smaller stones as well, leaving the surface both rough and porous. Overflows from blocked gutters or choked cross-drains cause much damage in the same way. The amount of material lost from the roads by this means is often larger than the toll exacted by traffic.

If an excess of water is detrimental to a road, however, an extended drought is little or no better. The winds remove the binding material both by blowing it directly from the surface of the road and by carrying off the dust raised by traffic. This causes the road to "ravel." These loose stones should be removed, as leaving them on the road not only makes traffic disagreeable, but also tends to loosen still others. The stones which are picked off the road will rarely prove of value for repair work, as they are too much rounded to bond readily.

The wheels of passing vehicles produce on the road several effects, which should be understood. First, there is the grinding and crushing action on the surface, and second, the pressure throughout the entire body of the road covering. If for any reason the materials are not thoroughly consolidated, there is a third action of displacement accompanied by internal wear as the stones rub against each other. If the road surface is hard, smooth and waterproof, the wear will be the least possible and will be confined to the surface. The aim should, therefore, be directed toward a constant maintenance of a hard, smooth, waterproof surface.

The actual amount of wear on any given road surface depends on several conditions, viz., the amount and kind of traffic, climate and other local conditions, and the kind of road material used. Generally speaking, however, the amount of wear is less in proportion as the road is kept in good condition as to surface, solidity, and drainage. It is usually less on slight grades than on dead level, because of the better drainage, but on steep hills it is increased by the effects of running water. Strange to say a hill usually looks better after a heavy rain than the flat below. This is because the hill is washed

MAINTENANCE AND REPAIR 197

clean, while the flat is more or less covered with the debris and mud carried down from above. This often leads to a neglect of the hills until they are so badly worn as to require resurfacing.

The amount of maintenance required will vary with the season and the local conditions. Ordinarily there is more wear in winter than summer, and more in wet places than in dry. The reverse, however, is true on roads with heavy automobile traffic, when dry weather proves especially injurious.

The amount of wear is also greatly augmented by the prevailing tendency of the traffic to follow in the same track, especially where the surface is soft, so that the tracks become visible. In parts of Germany the road labourers have a custom of placing large stones on to the road whenever a rut or depression tends to form, because of the concentration of the traffic along this one line. The stones serve to deflect the traffic and so keep the wear uniformly distributed over the entire road surface.

Proper maintenance consists in replacing the materials lost by unequal wear, so that the road is always in a good smooth condition. In making repairs, the materials should be spread only on the places which require them. Thus no portion of the surface is neglected, and no materials are wastefully applied to portions already thick enough to stand the traffic. Uniformity in both strength and smoothness with the least use of materials and least cost is the thing to be sought. In spreading new stone, the old method of waiting until the road has entirely lost its shape and then spreading a thick coat which is left to be worked in wholly by the traffic cannot be too severely condemned. This method is very wasteful of material, as well as extremely inconvenient to traffic. A great deal of the material is ground up and crushed before it is consolidated, and even after consolidation the surface is rarely if ever left smooth. The materials necessary to replace the loss by wear of ordinary traffic should be spread in comparatively small quantities where hollows or weak places occur, or where required to keep the cross section of the road in proper form. If laid in with care and in small patches, the inconvenience to traffic will scarcely be notice-

MAINTENANCE AND REPAIR

able. If the task of consolidating the materials laid is to be left to the public, it is only proper that they should demand that the process be made as easy and speedy as possible, which is readily attained by good arrangement and care in spreading the materials, and close attention afterwards until they are consolidated. Where, because of neglect, or other reasons, it is necessary to make extensive repairs or resurfacing, the steam roller should always be employed to do the consolidation. Especially is this true if the road is extensively used by automobiles.

The steam roller is also a very useful machine in the maintenance of roads softened by winter frosts. A few trips in the spring of the year soon after the frost has left the ground will remove the slight ruts beginning to form, and recompact the road's surface, rendering it hard, smooth and waterproof. All loose stones should be removed before the road is rolled.

The following instructions to road men, issued by The Road Improvement Association of London, should be found useful in the maintenance and repair of macadam roads:

1. Never allow a hollow, a rut, or a puddle to remain on a road, but fill it up at once with chips from the stone-heap.

2. Always use chips for patching, and for all repairs during the summer months.

3. Never put fresh stones on the road, if by crosspicking and a thorough use of the rake the surface can be made smooth and kept at the proper strength and section.

4. Remember that the rake is the most useful tool in your collection, and that it should be kept close at hand the whole year round.

5. Do not spread large patches of stone over the whole width of the road, but coat the middle or horse track first, and when this has worn in, coat each of the sides in turn.

6. Always arrange that the bulk of the stones may be laid down before Christmas.

7. In moderately dry weather and on hard roads, always pick up the old surface into ridges six inches apart, and remove all large and projecting stones before applying a new coating.

8. Never spread stones more than one stone deep, but add a second layer when the first has worn in, if one coat be not enough.

9. Use a steel-pronged fork to load the materials at the stone-heap, so that the siftings may be available for "binding" and for summer repairs.

10. Go over the whole of the new coating every day or two with the rake, and never leave the stones in ridges.

MAINTENANCE AND REPAIR 201

11. Remove all large stones, blocks of wood, and other obstructions (used for diverting the traffic) at nightfall, or the consequences may be serious.

12. Never put a stone upon the road for repairing purposes that will not pass freely in every direction through a 2-inch ring and remember that still smaller stones should be used for patching and for all slight repairs.

13. Recollect that hard stone should be broken to a finer gauge than soft, but that the 2-inch gauge is the largest that should be employed under any circumstances where no steam roller is employed.

14. Use chips, if possible, for binding newly laid stones together, and remember that road-sweepings, horse-droppings, sods of grass, and other rubbish, when used for this purpose, will ruin the best road ever constructed.

15. Remember that water-worn or rounded stones should never be used upon steep gradients, or they will fail to bind together.

16. Never allow dust or mud to lie on the surface of the road, for either of these will double the cost of maintenance.

17. Recollect that dust becomes mud at the first shower, and that mud forms a wet blanket which will keep a road in a filthy condition for weeks at a time, instead of allowing it to dry in a few hours.

18. See that all sweepings and scrapings are put into heaps and carted away immediately.

19. Remember that the middle of the road should

always be a little higher than the sides, so that the rain may run into the side gutters at once.

20. Never allow the water-tables, gutters and ditches to clog up, but keep them clear the whole year through.

21. Always be upon your road in wet weather, and at once fill up with "chips" any hollows or ruts where the rain may lie.

22. When the main coatings of stone have worn in, go over the whole road, and gather together all the loose stones, for loose stones are a source of danger and annoyance and should never be allowed to lie on any road.

The Problem of the Automobile.—In the last few years the need of proper maintenance and possibly even a radical departure from some of the former methods of maintenance, especially on roads near large cities and the principal thoroughfares between cities, has been greatly emphasised by the advent of the automobile. We are confronted by a dust problem due to this new vehicle. Dust has always existed. The chemical, physical and mechanical agencies which produce the dust are in no way new. The automobile, when not equipped with chain tires, is not a dust producer in that it grinds up the road material; it takes the dust made by other agencies and disseminates it over the surrounding country. The broad-tired, swiftly moving automobile throws the dust from between the stones and the strong, deflected wind current from the car blows the dust from the road surface into the air to be carried away by the wind to the detriment of the road, the travellers, nearby residents and bordering foliage.

No one will seriously question the statement that the automobile has come to stay. Nor will it be wise, even though it should prove possible, to limit the speed below that consistent with the proper safety of all concerned. The solution must be found, either by a change in the design of the cars so as to raise less dust, or by the highway engineer in the construction and maintenance of the roads in such a manner as to prevent the formation of the dust, or so as to retain it on the roadway when formed. Probably the ultimate solution will come from both sources. The other part of the solution rests with the highway engineer. Present indications point to two lines of procedure: preventing the formation of dust, and laying the dust when formed.

The prevention of the formation of dust includes the selection and use of materials which give very little dust, that is, those which bond very well and are very resistant to abrasion, and second, the use of binding materials other than rock dust in road construction. One of the evil effects of the automobile traffic at present is that the binder is all blown away, leaving the surface free to ravel, which in turn produces more dust to be blown away by the automobile. By using only the best materials, the dust nuisance can be lessened to a considerable extent. Before a dust preventive of any kind is applied, however, the road must be in good condition, i.e., good repair. Dust preventives are simply another step in road maintenance, and in no way vitiate the need for a smooth, properly drained and properly repaired road surface. Having secured this, you are ready to take the advance step of applying some dust-layer or surface dressing to prevent its rapid formation; in other words, to lessen the wear on the road, for much dust usually means that the road surface is wearing rapidly. The dust on the road has

MAINTENANCE AND REPAIR 205

but two sources, viz., the foreign material brought on and ground up, such as horse droppings, etc., and the material abraded from the road surface. It is the latter which chiefly concerns the highway engineer. The rate at which it is formed is in a manner a measure of the wear of the road.

The evil effects however, go much further than the mere destruction of the road surface. Travel for health and pleasure is practically prohibited by the thick clouds which fog up from the disintegrating surface. This dust is carried by winds to the neighbouring fields and houses, to the extreme annoyance of the roadside dwellers and to the detriment of the crops and foliage along the way. In not a few places, the values of otherwise desirable properties have declined greatly because of the dust which, in extreme cases, prohibits the use of front porches and open doors or windows on the side toward the road. The question of public health is even a more vital one, however. Dust and disease are most intimately connected. Tyndall, the great scientist, once declared that the ravages of war are small compared to the victims claimed by that insidious, relentless arch enemy of mankind, dust.

The proper use of the various substances which are used as binders or dust layers is discussed in the chapter on Modern Road Problems.

CHAPTER X

ROADSIDE TREATMENT

ROADSIDE treatment has received comparatively little attention in the United States, and yet proper attention to the roadside is not only essential to the beauty of the road and to the pleasure and comfort of the travellers, but also to the preservation of the road itself.

Roadsides.—After a road is completed, rubbish should be removed, and excavations and embankments, except such as are necessary to the road, should be smoothed over and sown with grass, and all unsightly brush and weeds removed. In short, wherever possible, the road should run between strips of smooth green sward, and suitable shade trees should be planted at intervals, so as to provide a pleasing appearance, shade for the traveller, and protection to the road from drying out too rapidly, provided it is macadam or gravel. Clay and earth roads should be free of shade. Shade trees are an important factor in reducing the cost of maintenance of macadam roads, by reason of the fact that they prevent the road from drying out and becoming dusty.

In the selection of shade trees care should be taken to secure only those which are suitable to local conditions. In all cases it is well to choose a tree that is hardy, grows rapidly, and has abundant foliage. A good plan is to plant trees with tops fifty feet apart, but alternating on each side of the road, so that there will be a tree every twenty-five feet. In some portions of Germany fruit trees are planted extensively along the roadside, and a considerable revenue is derived from the sale of fruit. In Saxony apple, pear and cherry trees are planted along the road from 90 to 120 feet apart, and plum trees about 25 feet apart. Upwards of $21,000 a year has been obtained from the State roads of Saxony from this source, and still larger amounts from local roads. In India the Government allows abutting property owners to take the produce of fruit trees in exchange for protecting and caring for the trees. The irrepressible American boy is a factor which would have to be taken into consideration, if such a plan were ever contemplated in this country. With our present inadequate system of maintenance which does not provide for daily patrol, it would probably be better to resort to forestry rather than to horticulture for guidance in roadside tree planting.

Effect of Trees on Roads.—The beneficial effect which is most generally apparent from the planting of trees is the prevention of dust in summer. On the other hand, it is contended that they prevent muddy roads from drying out. The presence of trees along the roadside is generally a partial preventive of damage to the road from hard, driving rains. A road shaded by trees is cooler by day and warmer by night during the summer, and is warmer both day and night in winter. By preventing the loss of heat by radiation, trees and tall hedges reduce the freezing of the road surface and, consequently, protect the road in a measure against the destructive action of frost. Shade also prevents the destructive effect due to rapid thawing of the road by strong sunshine in the spring. A great deal of damage is done to unshaded roads by traffic passing over them while the rapid thawing process is going on.

Protection From Wind and Snow.—When determining upon the kind of roadside treatment to be adopted, consideration should be given to the protection of the road from snowdrifts in sections of country where the snowfall is heavy. A study of the relative positions of snowdrifts, the direction and velocity of winds, and the relative location of the road, would aid in determining what course to pursue; for example, whether trees or hedges would be most advisable, and if trees, what kind should be used, or if hedges, the kind, height, location, and method of planting.

The protection of stone and gravel roads from wind is very important, as the continued prevalence of high winds tend to strip the road surface of the rock dust which is essential to the bond of the road. The injurious effect from wind is most pronounced in summer when the roads are dry. Consequently, if the roadside is planted with trees or hedges, the foliage will be thickest in summer, so as to afford a screen which will materially lessen the force of the wind before it reaches the road surface.

The Kind of Tree to Select.—As previously stated, the important considerations in the selection of roadside trees are: first, adaptability to local conditions; second, hardiness; third, good foliage; fourth, rapid growth. Wherever practicable, trees of local origin should be used.

There are a great variety of conditions in the United States, and it would be impossible to designate a list of trees which would be adaptable to all the road conditions which might exist in this country, unless it were desirable to limit the list to fruit or nut-bearing trees. If this were the case, the fruit-bearing trees which would be best adapted to road conditions would be the apple, and possibly the pear, in some localities. Apples would cover all that section of the eastern United States north of the Carolinas, and even south of this in the Appalachian region. West of the mountains the apple would serve as far south as the Gulf States, and west to the base of the Rocky Mountains, with perhaps the exception of the extreme northern part of Minnesota, the Dakotas and Montana, where some other plants would have to be substituted for the apple, unless the crab were used. The nut-bearing trees which are adapted to this use in the eastern United States are hickory, walnut and butternut for the New England States, and along the Appalachian Mountains as far south as Georgia; but the distribution of these nut trees would take a northern turn on the west side of the Alleghany Mountains, and they should not be used, perhaps, south of central Kentucky, and no further west than Colorado. The hickory will not thrive in northern Iowa, northern Wisconsin, Minnesota or the Dakotas. The black walnut, however, will grow well as far north as the southern part of Minnesota, over the eastern part of South Dakota, eastern Nebraska and Kansas. On the Pacific Coast the English walnut can be used as a substitute for the nut trees grown in the eastern part of the United States, and in the South Atlantic States and the Gulf States pecans may serve as a substitute for the other nut trees mentioned.

Ordinarily it is better to select some long-lived shade tree than to attempt to combine fruit production with shade. For the New England and Middle States the sugar maple is one of the most extensively used and one of the most desirable shade trees for this purpose. Elm is very desirable, but it does not produce as dense a canopy as the maples. If a more dense shade is desired than that produced by the sugar maple, the Norway maple may be substituted. In localities from Washington, D. C., southward to the Carolinas, a variety of shade trees may be employed, such as silver maple, which is perhaps the least desirable of all; the elm and red oak, similar to the varieties growing on Twelfth Street, Washington, D. C., along the side of the Smithsonian and Department of Agriculture grounds; the willow oak, a fine example of which is standing just across from the Henry Monument in the Smithsonian grounds, Washington, D. C.; the Norway maple, which has long been considered as one of the finest shade trees for that locality; the pin oak, which is being so extensively used on the streets of Washington; and the sycamore, which has a natural distribution throughout the Middle States. After the confines of the Carolinas have been reached, there is nothing which compares with the live oak. This should be planted to the exclusion of everything else throughout the southern part of the United States, because it is typical of the region and is one of the most beautiful trees grown in America. For California probably the pepper tree will supersede everything else as a roadside tree, while in Florida, the camphor tree might well be used as a substitute for the pepper tree in California. In extreme southern Texas the native palm could be used very effectively for roadside decoration. Where this is not desirable, the hackberry, both native and Mexican varieties, may be used to good advantage. For the extreme Northwest, including the Dakotas and northern Minnesota, perhaps the best roadside tree would be the American elm or the green ash.

CHAPTER XI

MODERN ROAD PROBLEMS

On roads which are subjected to heavy automobile traffic, the most important problem confronting highway engineers is the prevention of dust and the preservation of the road from the destructive action of automobiles moving at high rates of speed. The standard macadam road has been found inadequate to withstand this new form of traffic, especially when the automobile traffic is dense.

As previously explained in this volume, the macadam road was designed for the purpose of withstanding the wear of iron-tired horse-vehicles. On a macadam road properly constructed with suitable material, the amount of dust worn from the rock fragments is only sufficient to replace that which is carried away by wind and rain, so that the bond of the road is continuously preserved. The advent of the automobile has brought about new conditions.

The driving wheels of motor cars moving at high rates of speed exert a powerful tractive force on the road surface, which displaces the materials composing the surface. The result is that the finer particles and dust are thrown into the air to be carried off the road by cross currents of air. The rubber tire of the automobile does not wear any appreciable amount of dust from the rock fragments, and consequently, the loss of the rock dust is a permanent loss to the road. Under these conditions, the road soon ravels, making travel difficult and allowing water to make its way to the earth subgrade or foundation.

In spite of the fact that the automobile is responsible for these and other unfortunate conditions, it must be realised that the automobile is one of the most useful inventions of the age, as it brings distant communities in closer touch, and places at the disposal of man a power for the transportation of himself and his products of infinitely greater possibilities than animal power. The automobile has come to stay, and we could no more legislate it out of existence than we could abolish the railroad and the locomotive. At the present time it is estimated that from 12,000 to 13,000 automobiles are manufactured every month, and the number is constantly increasing. Up to the present time most of the manufacturers have devoted their energies to supplying the demand for passenger cars, but the time will undoubtedly come when the automobile will be used quite generally for the transportation of farm products to market over good roads. It is necessary, therefore, that attention be directed toward providing roadways suitable for this new form of traffic.

Methods of Meeting Conditions.—All remedies which have been tried or suggested in this connection may be considered in two classes: first, those which deal with the construction of new roads, so as to minimise the formation of dust, and second, those which deal with the treatment of the surfaces of existing roads, to bring about the same results. In the construction of new roads various bituminous binders have been employed with crushed stone, and this type of road is known as the bituminous macadam. In the treatment of old roads various bituminous and other binders have been applied to the surface, according to a number of different methods. The materials which are applied to roads for the purpose of preventing the formation of dust may be considered in two classes: first, those that are applied in their original condition, and second, those that are applied in emulsion or solution in water.

Mineral Oil.—This material has been quite generally used in the treatment of road surfaces, with varying success. The oils that have given the most satisfactory results are those having an asphalt base. Asphalt forms an excellent binder, while paraffin has practically no binding power, and would merely result in making the road greasy. The eastern oils contain almost a pure paraffin base. Some of the Kentucky oils, and most of those in Texas, have a mixed paraffin and asphalt base, while many of the California oils contain a high percentage of asphalt. The oil is applied either in the crude state or after distillation at refineries, where the lighter and more volatile parts are removed. An oil which has been refined in this way is known as a residual oil, is heavier and thicker than in its original state, and possesses a larger percentage of the base. Consequently, it is, as a rule, better suited for road treatment. When the oil is not too heavy, it can be applied to the road surface with an ordinary sprinkling cart, but when it is too heavy for use in this way, it is usual to heat it and apply it to the road by means of a sprayer, either with or without pressure. In the application of surface binders the best practice is to sweep the road clean, so that the binder may penetrate and be incorporated in the body of the road. After the material has been applied in this way, it is usual to place a thin covering of gravel, sand or rock screenings, and rock dust on the road. In California, oil has been applied to earth roads which have previously been ploughed up, and the materials thoroughly tamped and mixed by means of a tamping or sheep-foot roller. This method of construction has not proved successful in the East.

Coal Tar.—Many engineers favour the use of coal tar for the prevention of dust and the preservation of roads, but one of the greatest difficulties is to obtain a universally good material from different producers. Coal tar is a thick, black liquid, obtained as a by-product from the distillation of coal during the manufacture of illuminating gas and coke. The base of coal tar gives its value as a road binder. This base, which is known as coal-tar pitch, corresponds to the asphalt base of oils. In the application of coal tar to roads, the dust should be removed and the tar applied in practically the same manner as oil. The refined tar is usually superior to the crude product, but is more expensive. The application of tar to a surface should be made only in dry, warm weather, and when the road surface is perfectly dry, as good results cannot be obtained otherwise. It would hold true, consequently, that the tar itself should not contain any water, as the road surface absorbs water more rapidly than other materials. It is almost always necessary to heat the tar before it can be applied to the road, and the method is usually to provide large iron kettles, equipped with portable fire boxes and mounted on wheels, or, where the tar is supplied in tank cars, the heating is done before the tar is removed from the tank. It is applied to the road either by means of a sprinkling device, or by hand sprinklers and spread with brooms. The tar should be allowed to dry for a few days before traffic is permitted on the road. The best practice is to spread a light course of sand or rock screenings over the surface after it has been treated with tar.

Solutions and Emulsions.—Materials other than tars, asphalts and heavy oils, are generally included under the term "palliatives." Previous to the introduction of motor vehicles, water was the agency generally relied upon to keep the dust down on stone roads. Owing to the fact that water evaporates rapidly, a number of chemical salts, having the property of absorbing and retaining moisture in the atmosphere, have been used.

Calcium Chloride.—Probably the best example of this form of dust preventive is known as calcium chloride, a by-product produced in the manufacture of soda. Calcium chloride has a great affinity for water, and absorbs and retains moisture from the atmosphere for a considerable length of time. It is, however, only temporary in its effect, as compared with the heavier binders. It is prepared for application to the road by mixing with water, and is applied by means of the ordinary sprinkling cart.

Waste Sulphite Liquor.—A waste product from the wood-pulp paper-mills has been recently used with some success, but as the base of this material is soluble in water, it can be classed only as a temporary binder. The best results have been obtained from this material by the application of a concentrated solution of about 1.13 specific gravity, at the rate of ½ gallon per square yard. Under favourable conditions, this treatment will keep the dust down for a whole season, and the material may, therefore, be considered as a semi-permanent binder.

Permanent Binders.—Coming now to the permanent binders, we may consider their use according to two general methods of construction, known as the penetration or grouting method, and the mixing method. Among the permanent bituminous binders which we have so far employed are the heavier residual oils and tars of semi-solid or solid consistency, fluxed oil and tar pitches and solid native bitumens, and fluid cut-back products, which are capable of increasing in consistency after application, by volatilisation of the lighter constituents.

When employing the penetration method, the best practice is to construct the road as follows: Upon the subgrade, prepared as for ordinary macadam work, a foundation course of No. 1 crushed stone is placed to the desired depth and well rolled. Sufficient screenings are then applied to fill the surface voids, and care is taken that there is no excess of fine material which will prevent the wearing course from keying into the foundation course. The road is then rolled until absolutely firm, and more screenings are applied if necessary to take the place of those worked into the foundation. The wearing course of No. 2 crushed stone, clean and free from dust and screenings, is then applied to a finished depth of two or three inches and this course is lightly rolled. The hot bitumen is next poured or sprayed upon the road at the rate of from 1 to 1½ gallons per square yard, after which a light coat of clean ½-inch stone chips, free from dust, is applied and the road is well rolled. A seal coat of bitumen is then painted upon the surface at the rate of from a third to half a gallon per square yard, after which screenings are applied and rolled in until the road is smooth and firm.

Such a method of construction should produce a durable road on which dust formation is reduced to a minimum. The mixing method is, however, to be preferred because of the greater certainty of obtaining an absolutely uniform wearing surface in which each individual fragment is known to be covered with the binder.

In general the mixing method is conducted as follows: Upon a foundation course of crushed stone, prepared as just described, a mixture of crushed stone and bitumen is laid to a finished depth of from two to three inches, and rolled with the addition of screenings. A paint coat of bitumen is then applied and the road is finished in the manner previously described. The mixture of stone and bitumen may be prepared either by manual labour or machinery, preferably by the latter. The mineral aggregate may be graded in any approved manner. For country road work, the crusher run of stone from 2 inches or 1½ inches to dust may sometimes be satisfactorily used. In the experimental work of the Office of Public Roads a mixture of 27 parts crusher run of from 1½ inches to ¾ inch with 10 parts crusher run of from ¾ inch to dust has been found to produce a very dense aggregate. Such an aggregate should be mixed with not less than six per cent. of bitumen, and neither the stone nor bitumen should be heated above 350° C.

The application of rock asphalt in macadam-road construction may in a certain sense be considered as a combination of the penetration and mixing methods. This material, if containing a good grade of bitumen, will serve as a permanent binder. It has been mixed by nature so that each individual fragment is thoroughly coated with the binder. It is seldom suitable for use as a wearing surface of any considerable thickness, owing to the softness of the bitumen and to the fineness of the mineral particles, which are not, as a rule, well graded. If forced into the wearing course of a newly constructed macadam road, to which no screenings have been applied, it may, as has been demonstrated, prove to be a very serviceable road material which prevents excessive dust formation by reducing wear and disintegration.

In order that bituminous roads may be kept dustless it is necessary that they, in common with all other roads, be treated from time to time according to one of the first two methods mentioned earlier in this chapter, that is, they must either be scavenged, or their surfaces treated with temporary or semi-permanent binders. It will be found that the use of a good semi-permanent bituminous binder in comparatively small amounts will not only lay the dust satisfactorily upon these roads, but that such treatment will appreciably lengthen the life of the road by revivifying the old binder originally used during construction.

Portland cement is in some respects an almost ideal permanent road binder, especially when motor traffic only is encountered. When mixed in proper proportions with a suitable mineral aggregate it produces a hard rigid concrete, well designed to withstand the shearing strains exerted upon it by the driving wheels of automobiles, and practically unacted upon by the large pneumatic tires of such vehicles. Under steel-shod horse-drawn traffic it is, however, far from ideal, owing to its lack of resiliency and tendency to spall under impact and abrasion.

Investigations are now being conducted by the Office of Public Roads with a view to finding some way of overcoming these undesirable properties. While much experimental work will yet have to be done along this line, what has already been accomplished would seem to indicate that certain fluid petroleum residuums may be used to advantage in wet cement concrete mixtures, both for the purpose of waterproofing the concrete and reducing its tendency to spall. It has been found that there is little difficulty in incorporating petroleum residues with such mixtures, providing they are sufficiently fluid to be handled when cold, and some of the oils seem to produce no loss in the strength of the concrete in which they are used. We hope in the near future to try out some of these mixtures in the construction of experimental roads where it will be possible to study the results produced under actual service conditions. Until this is done it is, of course, impossible to draw any definite conclusions in regard to the practical value of mineral oils as road preservatives and dust preventives in the construction of cement concrete pavements.

In conclusion a word may be said in regard to the value of an oil binder in earth-road construction. While the experiments conducted by the Office along this line have been in no sense failures, it would seem that in the eastern part of the United States at least, the oiled earth road is not destined to prove a success. This is due both to the character of the oils which have to be used and to the rather severe climatic conditions encountered here.

CHAPTER XII

PATHS

THERE are many sections of the country where the roads are so poor as to render them practically impassable for pedestrians, especially during the winter months. In other parts of the country where the roads are improved, the automobile and wagon traffic is very frequently so heavy as to render the roads not only disagreeable but dangerous to travel on foot. School attendance in many rural districts is sometimes interfered with, and in some instances the schools have to be closed on account of bad roads for varying periods during the winter months.

With the expenditure of a comparatively small amount of money, sidewalks or side paths could be built to accommodate pedestrians along the main roads where the conditions above described prevail. Side paths would facilitate school attendance and at the same time encourage the healthful exercise of walking, a pastime that is too seldom indulged in by the average American.

Paths should be located on the highest side of the road on the slope or shoulder just outside of the surface ditch. A strip of sod a few feet in width should be provided between the path and the road, as otherwise it will be found, if the path is on the same level as the road, that teamsters will drive their heavy wagons and horses over the line of the path and destroy it. The width of such paths need not be greater than 2 or 3 feet.

A Sand-Clay Path.—For all ordinary purposes, a path built out of a mixture of sand and clay, or out of fine gravel, will serve every purpose. No foundation will usually be required, as the paths are not subject to heavy loads. If the ground over which the path is located is composed of clay or loam, and the surfacing is to be made with sand and clay, or sand and loam, the ground should be ploughed up slightly and then covered with a thin layer of sand. This sand should be mixed with loam or clay by means of a garden rake. The mixing should be done when the soil is comparatively damp, and the mixture should contain from 85 to 90 per cent. of sand and from 10 to 15 per cent. of clay or loam. After the mixing process has been completed, a thin layer of sand should be applied to keep the clay or loam from becoming muddy or sticky in wet weather. The path should be slightly crowned so as to shed surface water, and small tile culverts should be placed under it at low places, so as to prevent washing. It will be unnecessary, as a general rule, to construct bridges and culverts for the paths, as the ordinary bridges and culverts of the roadway can be used by pedestrians.

It will sometimes be found that the surface soil by the roadside contains about the right mixture of sand, gravel and clay to make a good path, and under such circumstances, all that is needed is to clear the right of way and remove weeds, rocks and other obstructions, and crown the surface. Where gravel is used, all large pebbles should be raked out and discarded or used for the foundation. There is nothing quite so disagreeable as to walk on a road where the surface is covered with large stones or pebbles.

Stone Screenings, Cinders, etc.—Cinders are very frequently used for side paths, but as they are lighter and more friable than gravel, sand or crushed-stone screenings, they are more liable to wash, and consequently do not give as satisfactory results. Crushed sandstone, limestone chips and screenings, or other screenings, make good materials for paths, provided no piece is larger than ½ or ¾ inch in diameter, and that the mixture contains enough fine material to cover the coarser screenings. As a general rule, 2 or 3 inches of any of the materials above mentioned will be sufficient. The materials may be laid directly on the sod or soil, as this insures better drainage than would be secured by digging a trench into which these materials are placed.

When side paths are built in thickly settled regions, or in the neighbourhood of towns and villages, it is often found desirable to construct them out of more permanent materials than those referred to above. The materials ordinarily used for this purpose are brick and concrete.

This advice applies in a general way to the

PATHS

construction of paths about one's house and grounds.

Brick Walks.—In building brick paths, an excavation of 4 or 5 inches should be made to the desired width of 3 or 4 feet. The foundation may be composed of cinders, gravel or crushed stone, placed to a depth of from 2 to 3 inches, and covered with a layer of sand to a depth of about 1 inch. A curb should then be laid on both sides of the walk, composed of bricks set on end with the upper ends flush with the surface of the ground. The bricks for the walk should be laid flat and not on edge. Every alternate brick should be laid lengthwise of the pavement, so as to avoid long cleavage lines or cracks. The bricks should then be tamped into position under a board, or, if possible, rolled with a light roller. The surface is then covered with a thin layer of fine sand which is broomed into the cracks.

Walks of Portland Cement.—The success of a cement sidewalk is largely dependent upon the provision of the following essential features of construction:

1. A firm, but porous foundation, to provide means for draining off rain water.
2. A sufficiently thick base of well-made, strong concrete.
3. A wearing coat of rich mortar, troweled to a smooth, dense surface.
4. The division of the walk into blocks, with lines of weakness between them, so that all cracks due to settlement, shrinkage or frost, will be made to occur at the joints, and will thus not be noticeable.

If proper drainage is not provided under a cement sidewalk, rainwater will accumulate, and in consequence the frost action will be severe in causing heaving in cold weather, and unequal settlement of the walk will occur when the ground is wet. Good drainage may be secured by laying a foundation of cinders, broken stone, gravel or coarse sand. Before laying the foundation, the ground is excavated to the proper depth and is well consolidated by ramming. The depth of the foundation course is dependent on the climate and the nature of the soil. In cold climates, or where the ground is soft, the foundation should be from 4 to 8 inches deep, while, in the more temperate climates, where no frost occurs and the soil is sandy and not likely to become soft or water-soaked, no foundation is required.

The foundation course should be thoroughly rammed to present a firm, unyielding surface, and if sand or cinders are used, they should be thoroughly wet when being compacted.

The main body of the concrete walk is made of coarse concrete and is called the base. The usual proportions for the cement base are one part of cement, two or three parts of sand, and five parts of broken stone or gravel.

Nothing but good Portland cement should be used in sidewalk construction. Natural cements are unsuitable, since they will not stand the wear, while Puzzolan cements are likely to suffer deterioration through the action of wear and weather.

The sand used in the concrete mixture should be clean and coarse; the stone should be hard and tough, and as free from dust as possible. If gravel is used, it should be thoroughly screened and should be free from clay or other matter likely to interfere with the proper adhesion of the mortar. Gravel as it comes from the banks should not be mixed with the cement to form concrete, but the sand should be screened out and recombined with the coarse particles in the desired proportion. The cement, sand and stone, or gravel, should be mixed with enough water to form a mixture of quaking or jelly-like consistency. Care should be taken to mix the materials thoroughly, so that each piece of stone will be coated with mortar.

Forms should be provided along the sides of the walk to confine the concrete. These may be made of wooden strips 1½ inches thick and of suitable depth, depending on the thickness of the concrete base. The strips should be nailed to wooden stakes, so that the tops are level with the finished surface of the walk.

It is necessary to lay the concrete base in blocks with definite lines of separation, so that in the event of settlement, shrinkage or temperature changes, the irregular cracks which would otherwise form will be made to occur in straight, well-defined lines. The separation into blocks may be made by steel plates or strips about ¼ inch thick, which are removed just before the final finish and joint is made. Another method is to lay the blocks alternately and fill in between them. Good results can be obtained by cutting the concrete course to a width of ½ inch, and finishing the top coat into the cut to the depth of 1 or 2 inches, and cutting with a trowel through both to separate the blocks when finished.

The size of the blocks depends upon the width of the walk. Blocks nearly square in shape have a better appearance than elongated blocks. The limit of size for a 4-inch walk is generally placed at 6 feet square.

The mortar wearing-surface should be placed as soon as a few of the concrete blocks have been placed, and before they have set. This surface consists of a mixture of cement and sand, cement and finely crushed stone, or cement and a mixture of sand and finely crushed stone. Care should be taken to proportion the materials exactly, and thoroughly mix them so that the surface will be of uniform colour throughout. The size of crushed stone usually specified is that which will pass a ¼-inch sieve.

The consistency of the mortar to be used is such as is ordinarily employed by a mason in laying brick. The mortar after being deposited on top of the concrete base is smoothed to the level of the side form by means of a straight edge guided by the top of the forms. The surface is then roughly floated with a plasterer's trowel, and soon after levelled with the straight edge. The final floating is not performed until the mortar has been in place from two to five hours, when it has partially set. For this operation a wooden float is first used and then a metal float, or plasterer's trowel. It is sometimes the custom to sprinkle a thin layer of "dryer," a dry mixture of 1:1 mortar, which is trowelled over the surface. This is not desirable, since it tends to make the walk glassy after floating.

The surface should be grooved. The mason locates the joints between the blocks of concrete by marks previously placed on the wooden side forms. The exact location of the joints is found by running a small trowel down into the joints in the concrete. By the use of a steel straight-edge the mortar coat is cut through in order to form the individual blocks. The corners of the cuts are rounded off by the use of

PATHS

a groover and edging trowel, which is a small float with one of its edges curved. A metal float is used over the entire surface to give it a final finish. To obtain a rough surface, a dotted or grooved roller may be employed.

It is advisable to protect the walk from the hot sun for several days after its completion; otherwise the surface is likely to dry out too quickly, with the consequent formation of shrinkage cracks.

The following table is compiled from the specifications for cement sidewalk construction, as practised in some of the larger cities throughout the country.

CONCRETE SIDEWALKS. REQUIREMENTS IN VARIOUS CITIES.*

City	Foundation		Base		Wearing Surface		Dry Coating	Size of Blocks	Guarantee in years.
	Material	Thickness	Thickness	Proportions	Thickness	Cement-Sand Proportions	Cement-Sand Proportions		
Boston	Broken stone, gravel or cinders	12	8	1:2:5	1	1:1		3½ to 6 ft. sq.	10
Rochester	Sand, gravel, broken stone or cinders	6		1:5	1	2:3			8
Philadelphia	Sand, gravel, broken brick, stone or cinders	8	8	1:2:5	2	1:2	1:1		
Washington		0	4	1:2:5	1	2:3	1:1		5
Chicago	Cinders or clean sand	12	4½	1:2:5	¾	1:1		5 ft. x 6 ft.	10
Milwaukee	Cinders or broken stone	4	2½	1:3:5	1	1:1		24 to 36 sq. ft.	
St. Louis	Cinders	8	3½	1:3	¾	1:1			1
Omaha	Gravel, slag or stone	4	3	1:2:4	1	1:2	3:1		5

* From Concrete, Plain and Reinforced, by Frederick W. Taylor and Sanford E. Thompson.

CHAPTER XIII

CULVERTS AND BRIDGES

BRIDGE construction is a very technical and highly specialized field of engineering in which the layman is not likely to make great headway. Nevertheless, there are many considerations regarding the construction and maintenance of culverts and small bridges which the layman can comprehend as well as the expert, and to these we desire especially to call the reader's attention.

By far the greater number of culverts and bridges on our public roads have a span of less than 50 feet. In the past these structures have, in general, been built of wood, but lumber so exposed is subject to rapid decay. Consequently, these structures require a great deal of repair and frequent renewals. The ever-increasing price of lumber is making the further use of wood for this class of structures more and more indefensible. The loads which our highway structures are called upon to sustain are also increasing. In many localities the movement of steam road-rollers and heavy traction engines is seriously hampered because of weak bridges and culverts. Thus, considerations of both economy and safety demand the use of other materials than wood in the construction of our culverts and bridges.

Road Bridges.—Road authorities should, in general, adopt some systematic plan of replacing all wooden structures as fast as they require renewal with permanent materials, such as concrete, and they should take particular pains to make sure that all new structures have sufficient strength to carry a heavy road-roller.

Wood should never be employed, except for very good reasons. When it is found necessary to use wood for the smaller culverts and bridges, certain practical, rather than theoretical, considerations should govern the builder. All beams and stringers should be carefully inspected and only those free from bad knots and other defects should be used. For flooring, select a lumber that is hard and tough. Planks that show a tendency to splinter should not be used. For 2- and 3-inch flooring, the spacing of the stringers should not exceed 2 feet in the clear. If the stringers are spaced as wide apart as the strength

CULVERTS AND BRIDGES

of the new plank permits, then in a short time when the plank is worn by traffic, failure will occur. Hence, to get the greatest possible amount of wear out of your flooring space the stringers closely.

Culverts.—Three prime requirements are necessary in the design of road culverts, viz., ample waterway, strength and durability. All culverts and bridges should be designed and built strong enough to carry safely the heaviest load which is ever likely to be hauled over the road. For short spans, this is usually a heavy steam road-roller. Many of the existing culverts and bridges are far too light to carry the loads which they should legitimately be required to carry. This is especially true where traction engines are numerous.

Durability is of the greatest economic importance. In many sections a large portion of the annual road levy is expended in repair and renewal of wooden culverts and minor bridges, and it is not unusual to find this practice defended on the grounds that the county or district cannot afford to build the higher-priced permanent culverts. This is simply a false sense of economy. True, the first cost of the permanent structure is greater, but there the outlay ends, while with wooden culverts, there is a large annual outlay for repair, as well as frequent renewals. Anyone interested in road improvements will find it most interesting to secure the following data for his own county or district: The number of culverts, cost of labour and material for repair and renewal each year, average life of wooden culvert, and the average life of wooden bridge floors. Then he could compute how long it would be before the actual present expenditure would pay for permanent culverts.

One of the frequent causes of the failure of culverts and bridges is due to inadequate waterway. Great care should be taken to provide a waterway ample to carry safely the largest storm-flow ever likely to occur. If the waterway is too small there is constant danger of a washout with interruption to traffic and high cost for repairs. On the other hand, if the waterway is made unnecessarily large, the cost of construction may be needlessly increased. Economical designs are those which provide adequately, but not extravagantly for all necessary requirements.

It is inadvisable to carry storm water any considerable distance along the road. Water, especially where the volume is ever likely to be large and the velocity high, is a grave source of danger. Every effort should be made to turn water away from the road before it gathers in sufficient volume to be dangerous. To lead water long distances along the road, so as to require but few culverts, is the poorest kind of economy, as well as faulty engineering. It is courting trouble and inviting disaster.

Pipe Culverts.—In many sections pipe culverts

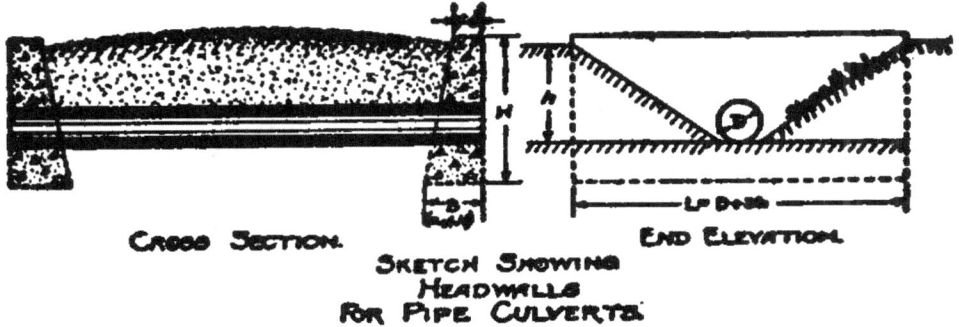

are proving very serviceable for sizes ranging from 12 to 24 inches in diameter. Because of the ease with which the smaller sizes of pipe clog and so become unserviceable, it is in general inadvisable to use for culverts sizes less than 12 inches in diameter, even though the amount of water to be removed could be carried by a smaller pipe. The kinds of pipe most commonly used are terra-cotta, concrete and iron.

Terra-cotta, or tile culverts, should be laid very carefully with the earth well tamped around them and provided with masonry or concrete end walls. There are three common causes for the failure of tile culverts: washouts, breakage by passing traffic, and breakage due to the expansive action of ice. These can, in general, be easily prevented. Washouts can be prevented by using a tile of proper size with careful placing, and the construction of suitable end walls—a point of great importance. End walls should be carried well below the pipe to a good foundation, and provision should be made against possible undermining by erosion.

If the soil is fine sand, or very friable, the joints should be laid in cement. When the soil is tough clay, hard pan or similar formation, this is not necessary, but in every case the earth

CULVERTS AND BRIDGES 247

should be carefully tamped beneath and around the tile.

To prevent breakage by passing traffic, it is in general only necessary to place the tile at a greater depth below the surface than has been customary. The most frequent causes of breakage is on earth roads, where the wheels of heavily loaded wagons cut deep ruts, sometimes actually striking the pipe, when, of course, failure takes place. The remedy is obvious. On earth roads, place the tile at such depth that wheels will not cut to or near it. To prevent the softening of the earth, provide good surface drainage. A load or two of gravel spread over the road at this point will also be of much assistance. In general, never place a tile culvert nearer than 18 inches to the surface on an earth road.

In cold climates, and especially in the prairie regions, tile culverts are often broken by the expansive action of the ice in winter. A tile culvert should never be placed where there is danger of the drainage being obstructed in such manner as to allow water or slush to accumulate in the pipe and then freeze. When the pipe is as much as two-thirds full of water

248   ROADS, PATHS AND BRIDGES

or slush, hard freezing will invariably burst it. For this reason, tile culverts must be used with caution in all cold countries, and especially in the flat prairie regions, where the natural drainage is poor.

Iron-Pipe Culverts.—With regard to iron-pipe culverts, the same care should be taken

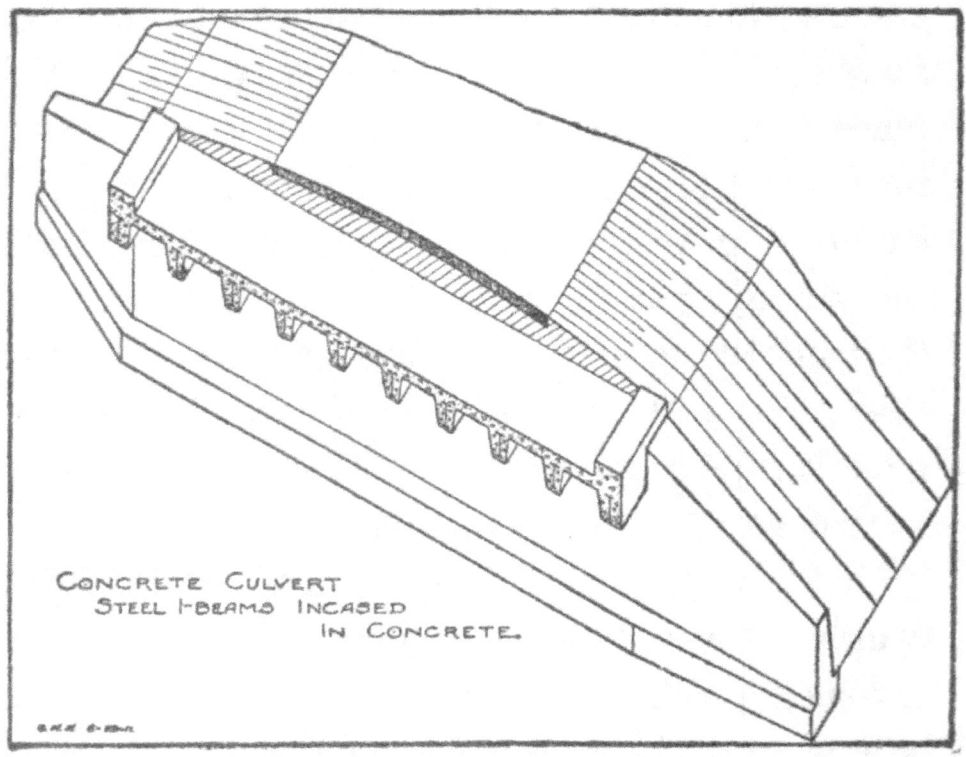

CONCRETE CULVERT
STEEL I-BEAMS INCASED
IN CONCRETE.

in laying and placing the end walls as with tile culverts. Improvements in the manufacture of iron have made this material more generally available for use in culvert construction. A special quality of iron, very low in carbon, is found to resist corrosion so well as to make its use advisable in many cases. The old style of cast-iron pipes is too heavy ever to come into general use. Corrugated iron pipe, however, when made from material of the proper quality, possesses strength together with durability and lightness. Corrugated iron pipe can be laid with somewhat less covering than tile pipe, and will successfully resist the expansive action of ice. It can, therefore, be used in places where it would be folly to place tile.

Concrete-Pipe Culverts.—Concrete, both plain and reinforced, is used to some extent in the manufacture of culvert pipes. When carefully made of proper materials, they are very serviceable. In general, it may be said that for the use of plain concrete culvert pipe, the same considerations govern as for the tile pipes, while the reinforced concrete culverts may be used wherever under other considerations the corrugated iron pipe could be used.

End Walls.—What the foundation is to a house the end walls are to a culvert. Without suitable end walls, a culvert is without protection and is placed in danger at every severe storm. Where the fill is high, wing walls may be used to hold it back, but for most pipe culverts they are not needed. In friable soils, and whenever the velocity of the water is high, the space at the outlet end of the pipe should be paved to prevent erosion and danger of undermining the end walls. The end walls should be carried down to soil sufficiently firm to prevent any settlement.

The materials for end walls may be brick, stone or concrete. If bricks are used, they should be hard burned and laid in cement mortar. Concrete is in general the best material for use in the construction of wing and end walls. The concrete should be about a $1:2\frac{1}{2}:5$ mixture The length of end wall should be about $D + 3H$, where D equals diameter of pipe and H equals height of fill above bottom of pipe.

Concrete Culverts.—In general the best material for use in the construction of culverts and the smaller bridges is reinforced concrete. The first cost of a reinforced concrete structure is naturally higher than that for a wooden one, but if properly built in the first place, the struc-

CULVERTS AND BRIDGES

ture will be permanent and the items of repair and renewals will be eliminated. Safety, which is of the greatest importance, will also be secured from the outset.

There are four general types of concrete culverts and bridges, i. e., box, T-beam, I-beam

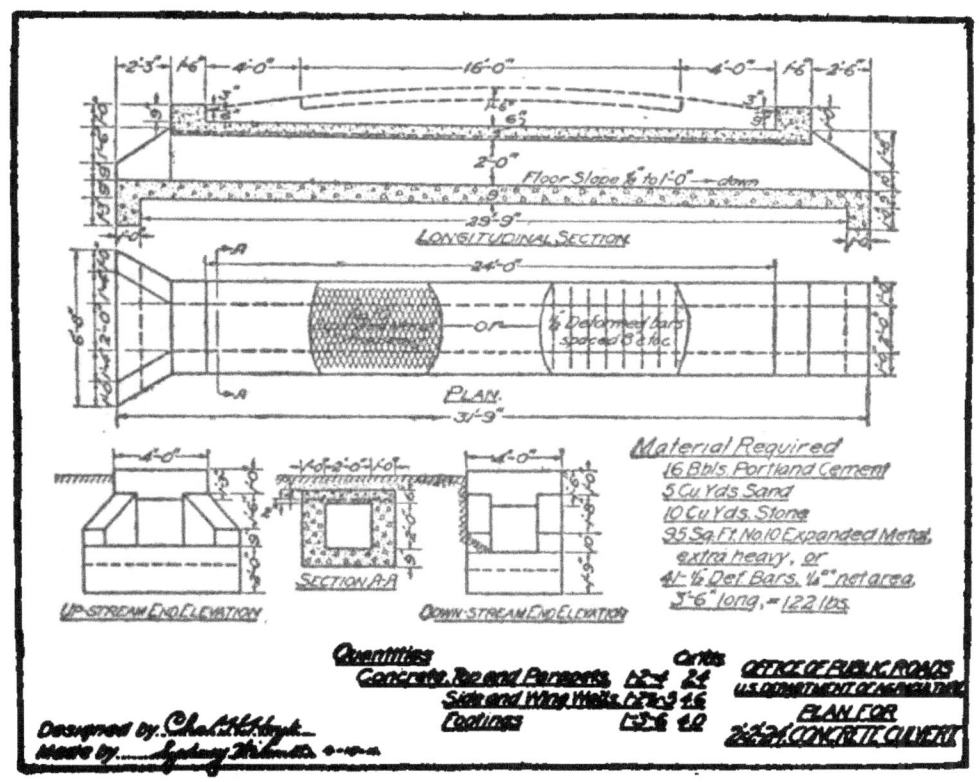

and arch. The box culvert may be used up to spans not to exceed 10 or 15 feet. The floor is a plain slab of reinforced concrete, while the abutments, wing walls and bottom may be built of either plain or reinforced concrete. Where it is not necessary to protect the foundation from erosion, or increase the heaving of the soil, the concrete bottom may be omitted or, in some places, a cobblestone paving may be economically substituted.

For spans above 10 or 15 feet, the T-beam type is quite generally used. Instead of increasing the thickness of the slab, the additional strength is secured by building longitudinal beams beneath the floor slab to carry the load. The beams and floor slab are built simultaneously, and steel reinforcement is placed near the bottom of the beam and also near the bottom of the floor slab. In the I-beam culvert steel I-beams are placed beneath the floor slab to carry the weight. The concrete covering of the beams is added simply to protect the steel from corrosion. This type of culvert is very readily constructed with but little skilled labour, and has in many sections proved very economical.

The arch culverts may be built of either plain or reinforced concrete, and are adapted for almost any length of span, provided the foundation is good and ample headroom may be had.

Arch culverts and bridges must be carefully designed and require close supervision during construction. Where the headroom is small, or the foundation poor, the arch type will, in general, not be found economical. Under no consideration, however, should this type of structure be attempted unless competent engineering assistance is available, both for the design and supervision of the construction.

Forms.—Much ingenuity and skill is required to secure economical forms which also provide for the necessary strength and tightness. The appearance of the concrete depends much upon how well the forms are made. Every crack between the boards and every joint that is poorly made or any other imperfection in the forms is filled with the wet concrete and leaves its impression upon the finished structure. For the boards next to the concrete, it is well to use green, or only partially seasoned, lumber, which is not so likely to warp and swell out of shape. For the parapets and other surfaces which will be exposed to view, it is advisable to use planed lumber. Working a spade or shovel along the boards will crowd back the larger particles, allowing the finer mortar to flow close to the boards, thus forming a smoother surface. The forms should be coated with soft soap or crude petroleum in order to prevent the concrete from adhering to them. On the parapets, wing walls, etc., the forms may be removed as soon as the concrete will safely hold its shape, and these exposed surfaces may be rubbed smooth with a wooden or brick block.

The main part of the forms should not be removed until the concrete has attained sufficient strength to carry safely the stresses to which it will be subjected. The time required will vary with the brand of cement used and the temperature. Concrete sets much more rapidly in warm than in cold weather. Ordinarily, the forms may be removed in from one to three weeks after the placing of the concrete.

Mixing and Placing Concrete.—In large quantities concrete is most economically mixed by machine. In smaller quantities, however, it is better to do the mixing by hand labour.

The following considerations should always be kept in mind. The materials should be so proportioned as to secure the densest mixture

CULVERTS AND BRIDGES

possible. The materials should then be so thoroughly mixed that each particle will be coated with a thin coat of mortar. In general work, for those parts requiring great strength, the proportion of 1:2:4, that is, by measure, one part cement, two parts sand, and four parts broken stone or gravel, gives satisfactory results. For hand-mixing a platform 10 or 12 feet square will be found advantageous. The sand is first placed on the platform and the cement added. The sand and cement are then turned with shovels until the mixture has a uniform colour. The stone or gravel is then placed on top, water is added with a bucket or hose, and the turning is continued until the whole is thoroughly and uniformly mixed.

Metal wheelbarrows are commonly used to convey material from the mixing platform to the forms. As the concrete is placed in the forms, it should be well tamped. The consistency of the concrete should be about such that, after thorough tamping, water should flush to the surface, and the concrete should have the appearance of quaking like a mass of jelly.

It is neither possible nor advisable in one

chapter to go into the details of bridge design. These belong to the highly specialized and technical field of bridge engineering. We have endeavoured only to cover those subjects for the proper understanding of which a high degree of specialized knowledge is not necessary. When it comes to structures of any considerable size, we can not be too emphatic in urging that competent engineering supervision be secured, both for the design and construction. Every question of both safety and economy in bridge construction has competent engineering supervision as a prerequisite. We know of no other way in which both safety and economy can be assured.

AUTHORITIES CONSULTED

In the preparation of this book the following authors were freely consulted.

AITKEN, THOMAS—Road Making and Maintenance, London, 1907.

BAKER, IRA O.—A Treatise on Roads and Pavements, New York, 1905.

BERGIER, NICHOLAS.—History of Great Highways of the Roman Empire, Brussels, 1728.

BLOODGOOD, S. DEWITT—A Treatise on Roads, Albany, N. Y., 1838.

BRUCE, P. A.—Economic History of Virginia in the Seventeenth Century, Vol. I.

BYRNE, AUSTIN T.—Highway Construction, New York, 1907.

COANE, JOHN MONTGOMERY—Australasian Roads, Melbourne, 1908.

ELLIOTT, CHARLES G.—Engineering for Land Drainage, New York, 1910.

FROST, HARWOOD—The Art of Road Making, New York, 1910.

GALLATIN, ALBERT.—Roads and Canals, Report to U. S. Senate, April 6, 1808.

GILLESPIE, W. M.—A Manual of Road Making, New York & Chicago, 1871.

GILLETTE HALBERT P.—The Economics of Road Construction, New York, 1906.

GREENWELL, ALLAN AND ELSDEN, J. V.—Roads, London, 1901.

HERSCHEL, CLEMENS—The Science of Road Making, New York, 1894.

HOOLEY, E. PURNELL—Management of Highways, London.

HUBBARD, PREVOST—Dust Preventives and Road Binders, New York, 1910.

HULBURT, ARCHER BUTLER—Historic Highways of America, 16 Vols., Cleveland, O., 1902.

JEFFREYS, REES—Dust Problem Statistics, London, 1909.

JENKS, JEREMIAH W.—Road Legislation for the American State, Baltimore, Md., 1889.

JOHNSON, J. B.—Engineering Contracts and Specifications, New York, 1902.

JUDSON, WILLIAM PIERSON—Road Preservation and Dust Prevention, New York, 1908.

LATHAM, FRANK—The Construction of Roads, London, 1903.

LOVEGROVE, E. J.—Attrition Tests of Road-Making Stones, London, 1906.

LOW, HENRY AND CLARK, D. K.—The Construction of Roads and Streets, London, 1901.

Preliminary Report of Inland Waterways Commission, U. S. Senate Document 325 60th Congress, 1st. Session.

RICHARDSON, CLIFFORD—The Modern Asphalt Pavement, New York, 1908.

RINGWALT, J. L.—Development of Transportation

AUTHORITIES CONSULTED

Systems in the United States, Philadelphia, Pa., 1888.

Ryves, Reginald—The King's Highway, London, 1908.

Searight, Thomas B.—The Old Pike, Uniontown, Pa., 1894.

Shaler, N. S.—American Highways, New York, 1896.

Spalding, Frederick Putnam—A Text Book on Roads and Pavements, New York, 1908.

Tillson, G. W.—Street Pavements and Paving Materials, New York, 1908.

Tucker, James Irwin—Contracts in Engineering, New York, 1910.

Annales Ponts et Chaussées, Paris.
Engineering and Contracting, Chicago.
Engineering News, New York.
Engineering Record, New York.
Good Roads, New York.
Surveyor, London.
Zeitschrift für Transportwesen und Strassenbau, Berlin.

Bulletins of U. S. Office of Public Roads.
Bulletins of U. S. Department of Agriculture.
Bulletins of U. S. Geological Survey.

THE END

INDEX

Appian Way, 10.
Asphalt, ancient use of, 5, 17. See BITUMEN.
Automobiles, effect on roads, 202, 215.
Authorities consulted, list of, 256.

Binders, permanent, 222.
Binding qualities of various stones, 169.
Bitumen, application of as a binder, 224.
Bond-issues for road-improvement advisable, 52.
Bridges, general considerations, 241.
Britain, Roman roads in, 12.
Broken-stone roads. See MACADAM.
Buckshot, or gumbo soil, 114.

Calcium chloride as a binder, 221.
Carthage, roads of, 8.
Causeway, Egyptian, built by Cheops, 4.
Cement walks, how made, 233.
Chariots of the ancients, 4.
Chert gravel, qualities of, 129.
Clays, properties of, 113, 116.
Clay road. See SAND AND CLAY ROAD.
Concrete walks, how built, 233.

Convict labour on highways, 46.
Crushers for stone, 147.
Culverts, construction and types of, 243–256; proper placing of, 94.
Cuts and fills, directions for, 98–101, 151.

Disc harrow and its use, 103.
Ditches, form and arrangement of 90, 105.
Drag, the split-log, form and use of, 183, 189.
Drainage of roads, importance of, 87; methods of insuring, 88–98, 152–155.
Dust, controlling the evil of, 204, 217, 226.
Dynamite, use of in road-making, 142.

Earth road, the, care of, 182; construction of, 79–109.
Economics of good vs. poor roads, 54–61.
England, early road-making in, 24; Turnpike Acts, 27.
Epaminondas of Thebes, 8.

Flaminian Way, 10.
France, origin of road system in, 20, 22.

Grade of a road defined, 82; method of determining, 85.
Gravel, qualities of, 125.
Gravel road, the, 124–133; care of, 187, 191.
Gumbo or buckshot soil, 114.

Highways, ancient imperial, 4, 5, 6–18.

Labor-supply affected by bad roads, 59.
Land, increase in value due to good roads, 57.
Legislation as to public roads, 22, 27, 38.
Loam, properties of on roads, 114.
Location of a road, advice as to, 64, 80; importance of, 62, 80; value of surveys for, 65.

McAdam, J. L., biography of, 28.
Macadam or broken-stone road, the, 134–162; care of, 194, 198; cost of, 161; courses of stone in, 157; drainage of, 152–155; first in the United States, 31; foundation for, 155; selecting materials for, 163–176.
Machinery and implements. See ROAD-MAKING MACHINERY.
Maintenance of roads, 177–236; American neglect of, 180; French system, 50, 180; importance of systematic attention, 49.
Mediæval ruin of the highways, 18, 21.
Mudholes, treatment of, 93.

National Turnpike, the, 33.

New England Path, the, 31.

Office of Public Roads, origin of the, 35; work of the, 36.
Oil for surfacing roads, 218.

Paths, construction of, 229–240.
Pavements in old times, 7, 25.
Personal service on roads a bad policy, 21, 44.
Peruvian road construction, 17.
Plans and specifications, rules for making, 68, 71–78.
Plough, the, and its use, 104.
Policy of road administration recommended, 40.
Population affected by condition of local roads, 59.
Post-service, ancient, 6; early posts in the United States, 31.
Public roads: public ownership necessary, 42; economics of, 54; financing, 57; more than local institutions, 48.

Quarrying for road-material, 139–143.

Repair of public roads, 177–206.
Roads in ancient times, 6–18.
Road-grader, the 103, 107, 148.
Road Improvement Association, rules of, 200.
Road machine, the, 107–109.
Road-making, revival of in Europe, 20.
Road-making machinery, 101, 143, 147.
Road roller, use of the, 147, 158.

INDEX

Road sprinkler, the, 147.
Roadside, treatment of the, 207–214.
Roadway, preparatory clearing of the, 98.
Rocks classified as road material, 165; specific gravity and weight of, 149; suitable for macadam, 163; tests for, 172.
Rock asphalt as a binder, 225.
Roman roads, construction of, 10–14; maintenance of, 14.

Sand and clay, directions for mixing, 115, 117.
Sand-clay road, the, 111–123; care of, 185.
Sandy roads, treatment of, 121.
School attendance affected by condition of the roads, 60.
Scrapers and their use, 102, 106.
Shade trees, selection of, 208, 211.
Snow, protection against, 210.
Spreader, the, 148.
Stone crushing, advice as to, 143–146.

Stone, first use in bridges, 5.
Subdrainage of roads, methods of, 94, 96.
Sulphite liquor as a binder, 222.
Surveys for new or improved roads, 65–78.

Tar for surfacing roads, 219.
Telearch, Greek, office of, 7.
Telford, T., biography of, 29.
Tile drains, how laid, 154.
Tolls, bad policy of, 22, 27, 33, 43; first collected, 6.
Trails, the forerunners of roads, 3.
Tresaguet, biography and work of, 23.
Turnpikes in the United States, 30, 33.

Walks of cement or concrete, 233.
Weeds, grass, etc., harmful, 109.
Width proper for roads, 98.
Wind, protection against, 210.

CONCRETE CULVERTS AND BRIDGES.

1. (Top.) Bridge of concrete on the State highway at Bucklin, Mass. 2. Tile culvert. 3. Arched culvert. 4. Culvert of the box type.

PATH OF STONE-SCREENINGS BESIDE AN OILED MACADAM ROAD (LITCHFIELD, CONN.).

THE AUTOMOBILE AND THE ROAD.

1. Motoring on a road of bituminous macadam. 2. Tearing up the pike. 3. How fast automobile travel affects a macadamized surface.

1. Modern traffic will not take the place of the road-roller. 2. A poorly built gravel-road in Illinois, with material heaped in the middle. "improvement." 3. Rocks hauled upon a road in Tennessee for its 4. Holes foolishly repaired with brush and weeds. 5. A corduroy road laid where gravel is abundant.

GOOD AND BAD MAINTENANCE.

1. (Top.) A French highway and one of its caretaking patrolmen. 2. An American example of worn-out macadam. 3. Rude road-mending in the southern United States.

EFFECT OF TREATMENT WITH A SPLIT-LOG DRAG.
A road in Iowa before and after dragging.

ROAD-MAKING MACHINERY.

1. Rolling the second course of macadam. 2. A portable stone crusher. 3. A spreader.

BAD ROAD-CONSTRUCTION.

1. (Top.) Loose stone thrown on and left to be packed by traffic 2. Raveling of macadam caused by use of stone which has little binding power. 3. Result of the use of quartz, lacking binding power; also, evidence of a bad foundation and poor drainage.

CONSTRUCTING A MACADAM ROAD.

The three courses of stone are shown in relative size: the largest ("No. 1s") at the bottom; the second, smaller ("No. 2s"), and the top or binder course of screenings; also a view of a road, showing its foundation, rolled, and a first course applied.

THREE SORTS OF GOOD ROAD.

1. (Top.) A well-constructed gravel road near Baker City, Oregon. 2. A road surfaced with slag screenings, to which quick lime was added as an additional cementing agent. 3. A sand-clay road near Aiken, S. C.

THE SAND-CLAY ROAD.

1. (Top.) Spreading the clay on the sand. 2. Plowing and mixing sand and clay. 3. Mixing with the disk-harrow. 4. (Left.) Method of laying a road-side drain.

ROAD-MACHINES HAULED BY A TRACTION-ENGINE.

AN EARTH ROAD WITH PROPER CROWN.

AN UNDRAINED PRAIRIE ROAD IN SPRING.

TRANSFORMATION OF AN EARTH ROAD.

1. (Top.) Present condition, improved by drainage and a macadam surface. 2. Past condition, sunken and water-soaked.

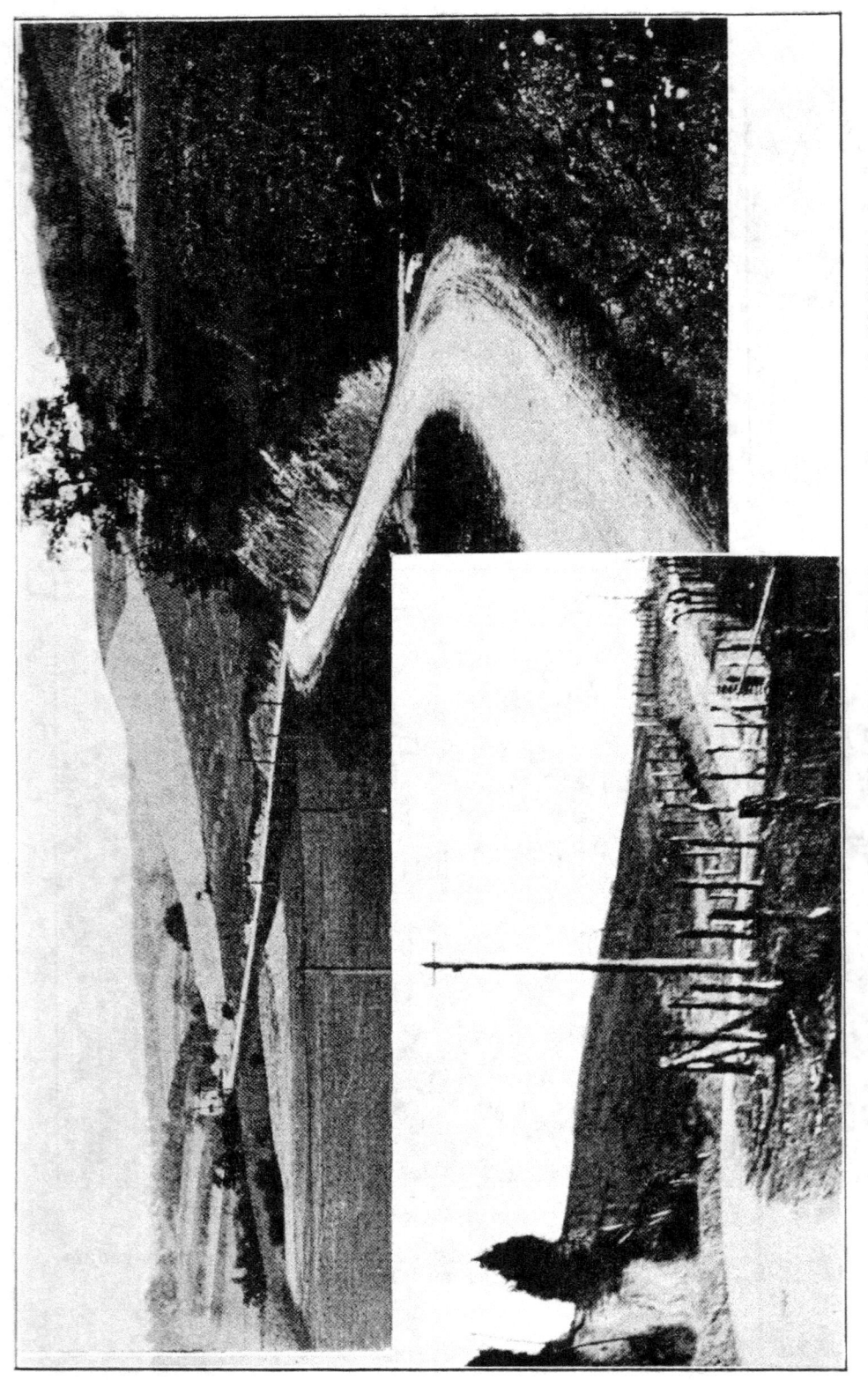

EXAMPLES OF GOOD AND BAD ROAD LOCATION IN A HILLY REGION.

THE ROADS AND THE SCHOOLS.

ECONOMICS OF GOOD-ROADS BUILDING.

Compare these heavy loads on a firm roadway, with this single bale taxing two mules in the mud.

DESTROYERS OF PROPERTY.

1. A mudhole on a road in a Virginia township which voted down a bond-issue. 2. A hill-road in another short-sighted community. 3. Ruined wagons about a blacksmith shop where roads are unimproved.

PRIMITIVE METHODS OF TRANSPORTATION.

A TOLL-HOUSE ON THE NATIONAL ROAD.

THOMAS TELFORD.

JOHN L. McADAM.

HOLLAND'S HIGHWAYS.

1. Taking milk to town over a country road in the Netherlands.
2. An interurban road in Holland, made of slag-brick; a cycle-path and shaded walk on the left and a bridle-path on the right.

SIMPLON PASS, SWITZERLAND.   PONT NAPOLEON.

A PAVED STREET IN POMPEII.

SOME ANCIENT HIGHWAYS.

1. (Top.) Appian Way and ruins of the Claudian Aqueduct. 2. Tombs along the Appian Way. 3. Avenue of Sphinxes at Karnak, Egypt.

www.ingramcontent.com/pod-product-compliance
Lightning Source LLC
Chambersburg PA
CBHW062318220526
45469CB00008B/2555